Managing Ethical Consumption in Tourism

Neither the tourism industry nor the tourist has responded convincingly to calls for more responsibility in tourism. Ethical consumption places pressure on travellers to manage a large number of decisions at one time when hedonic motivations threaten to override other priorities. Unsurprisingly, tensions occur and compromises are made. This book offers new insight into the motivations that influence tourists and their decision-making. It explores how consumers navigate the responsible tourism marketplace and provides a rich understanding of the challenges facing those seeking to encourage travellers to become more responsible.

Not only does this book provide an improved interpretation of the complexity of ethical consumption in tourism, it also offers a variety of stakeholders a deeper understanding of:

- the key challenges facing stakeholders in the production and consumption of responsible tourism
- how ethical consumers can be influenced to consume ethically
- the gaps in consumer knowledge and how to broaden the appeal for individuals to make more informed ethical decisions
- how tour operators can respond to this emerging market by innovative product development
- how to design informative marketing communications to encourage a greater uptake for responsible holidays
- how destinations can tailor their products to the ethical consumer market
- how destination communities and management organisations can target responsible tourists through the provision of sustainable alternatives to mass-market holiday products.

Written by leading academics from all over the world, this timely and important volume will be a valuable resource for undergraduate and postgraduate students, researchers and academics interested in Tourism Ethics, Ethical Consumption and the global issue of Sustainability.

Clare Weeden is a Senior Lecturer in Tourism and Marketing at the University of Brighton. Her research interests lie in the areas of ethical tourism, responsible tourist motivations, destination marketing and cruise tourism.

Karla Boluk is a Lecturer in Tourism at the University of Ulster. Her research interests lie in the areas of ethical/sustainable consumption in tourism, Fair Trade Tourism, FTTSA, social entrepreneurship and volunteer tourism.

Routledge critical studies in tourism, business and management

Series editors: Tim Coles
University of Exeter, UK
and
Michael Hall
University of Canterbury, New Zealand.

This ground-breaking monograph series deals directly with theoretical and conceptual issues at the interface between business, management and tourism studies. It incorporates research-generated, highly specialised cutting-edge studies of new and emergent themes, such as knowledge management and innovation, that affect the future business and management of tourism. The books in this series are conceptually challenging, empirically rigorous, creative, and, above all, capable of driving current thinking and unfolding debate in the business and management of tourism. This monograph series will appeal to researchers, academics and practitioners in the fields of tourism, business and management, and the social sciences.

Published titles:

Commercial Homes in Tourism (2009)
Edited by Paul Lynch, Alison J. McIntosh and Hazel Tucker

Sustainable Marketing of Cultural and Heritage Tourism (2010)
Deepak Chhabra

Economics of Sustainable Tourism (2010)
Fabio Cerina, Anil Markyanda and Michael McAleer

Tourism and Crisis (2013)
Gustav Visser and Sanette Ferreira

Managing Ethical Consumption in Tourism (2014)
Edited by Clare Weeden and Karla Boluk

The Routledge Critical Studies in Tourism, Business and Management monograph series builds on core concepts explored in the corresponding Routledge International Studies of Tourism, Business and Management book series. Series editors: Tim Coles, University of Exeter, UK; and Michael Hall, University of Canterbury, New Zealand.

Books in the series offer upper level undergraduates and masters students, comprehensive, thought-provoking yet accessible books that combine essential theory and international best practice on issues in the business and management of tourism such as HRM, entrepreneurship, service quality management, leadership, CSR, strategy, operations, branding and marketing.

Published titles:

International Business and Tourism (2008)
Tim Coles and Michael Hall

Carbon Management in Tourism (2010)
Stefan Gössling

Forthcoming titles:

Tourism and Social Marketing
Michael Hall

Managing Ethical Consumption in Tourism

Edited by Clare Weeden and
Karla Boluk

LONDON AND NEW YORK

First published 2014
by Routledge
2 Park Square, Milton Park, Abingdon, Oxon OX14 4RN

and by Routledge
711 Third Avenue, New York, NY 10017

Routledge is an imprint of the Taylor & Francis Group, an informa business

© 2014 editorial matter and selection: Clare Weeden and Karla Boluk; individual chapters: the contributors.

The right of Clare Weeden and Karla Boluk to be identified as the authors of the editorial material, and of the authors for their individual chapters, has been asserted in accordance with sections 77 and 78 of the Copyright, Designs and Patents Act 1988.

All rights reserved. No part of this book may be reprinted or reproduced or utilised in any form or by any electronic, mechanical, or other means, now known or hereafter invented, including photocopying and recording, or in any information storage or retrieval system, without permission in writing from the publishers.

Trademark notice: Product or corporate names may be trademarks or registered trademarks, and are used only for identification and explanation without intent to infringe.

British Library Cataloguing in Publication Data
A catalogue record for this book is available from the British Library

Library of Congress Cataloging in Publication Data
Managing Ethical Consumption in Tourism/edited by Clare Weeden and Karla Boluk.
  pages cm. – (Routledge Critical Studies in Tourism, Business and Management)
  Includes bibliographical references and index.
  1. Tourism–Moral and ethical aspects. 2. Tourism–Management.
  3. Consumption (Economics)–Moral and ethical aspects. I. Weeden, Clare.
  G155.A1M2617 2013
  910.68′4–dc23
                                                            2013026315

ISBN: 978-0-415-71676-5 (hbk)
ISBN: 978-1-315-87943-7 (ebk)

Typeset in Times New Roman
by Wearset Ltd, Boldon, Tyne and Wear

Printed and bound in the United States of America by Publishers Graphics, LLC on sustainably sourced paper.

Contents

List of figures	x
List of tables	xi
Notes on contributors	xii
Foreword	xv
Acknowledgements	xviii

1 Introduction: managing ethical consumption in tourism – compromises and tensions   1
CLARE WEEDEN AND KARLA BOLUK

PART I
Debates on ethical consumption in tourism   17

2 What does it mean to be good in tourism?   19
KELLEE CATON

3 You can check out anytime you like but you can never leave: can ethical consumption in tourism ever be sustainable?   32
C. MICHAEL HALL

4 Slow tourism: ethics, aesthetics and consumptive values   56
MICHAEL CLANCY

5 The evolution of environmental ethics: reflections on tourism consumption   70
ANDREW HOLDEN

PART II
Situating the self in ethical consumption 81

6 A fresh look into tourist consumption: is there hope for sustainability? An empirical study of Swedish tourists 83
ADRIANA BUDEANU AND TAREQ EMTAIRAH

7 Tourism's relationship with ethical food systems: fertile ground for research 104
CAROL KLINE, WHITNEY KNOLLENBERG AND CYNTHIA SHIRLEY DEALE

8 Travelling goods: global (self) development on sale 122
MARIA KOLETH

9 Exploring the ethical discourses presented by volunteer tourists 134
KARLA BOLUK AND VANIA RANJBAR

10 Ethical tourism: the role of emotion 153
SHEILA MALONE

PART III
Helping consumers make ethical decisions 167

11 Tread lightly through this *Lonely Planet*: examining ethical information in travel guidebooks 169
SARAH QUINLAN CUTLER

12 Business travel and the environment: the strains of travelling for work and the impact on travellers' pro-environmental *in situ* behaviour 188
WOUTER GEERTS

13 Medical tourism: consumptive practice, ethics and healthcare – the importance of subjective proximity 207
KIRSTEN LOVELOCK AND BRENT LOVELOCK

14 Marketing responsible tourism 225
CLARE WEEDEN

15 Concluding remarks 240
KARLA BOLUK AND CLARE WEEDEN

Index 248

Figures

7.1	Interface of tourism experiences and the food supply chain	105
11.1	Themes of ethical statements by guidebook	174
11.2	Themed statements showing ethical statement types	175
11.3	Guidebooks by ethical statement type	176

Tables

3.1	Approaches to consumer change	38
6.1	Number and percentage of respondents in Åre and Gotland, by socio-demographic category	87
6.2	Trip characteristics	88
6.3	Options for transportation, accommodation and leisure	90
6.4	Natural resource input per service unit for accommodation (overnight)	91
6.5	Categories of leisure activities based on energy intensity	91
6.6	Tourist choice of travel, accommodation and optional leisure	93
6.7	Environmental awareness	94
11.1	Coding scheme for statement-type categorization	172
11.2	Summary of ethical guidebook coding categories	173
11.3	Frequency and valid percentage for subject categories identified in ethical guidebook statements	178
11.4	Relating guidebook ethical subject categories to the principles of responsible tourism	179
12.1	Business travel interviewees	192

Contributors

Karla Boluk is Lecturer in the Ulster Business School at the University of Ulster. She has been equally engaged in social action throughout her academic career in a variety of international contexts (e.g. Canada, New Zealand, South Africa, Sweden, Ireland and the UK). Her main objective is to uncover the way in which tourism can be utilised as a catalyst for sustainable social change. Her research interests include: ethical/sustainable consumption, volunteer tourism, Fair Trade Tourism, FTTSA and social/societal entrepreneurship.

Adriana Budeanu is Associate Professor at Copenhagen Business School, Denmark, and has over ten years of experience in conducting research in the area of sustainable tourism and international business, sustainable consumption, corporate social responsibility in tourism supply chains, and sustainable service innovation. She has teaching experience in corporate strategies for sustainable development and sustainable tourism. Her most recent research studies tourist consumption and environmental awareness.

Kellee Caton is Assistant Professor in the Faculty of Adventure, Culinary Arts, and Tourism at Thompson Rivers University, Canada. She sits on the editorial board of the *Annals of Tourism Research*, the executive of the Tourism Education Futures Initiative and the scientific committee of the Critical Tourism Studies conference series. Her research interests include issues of morality and ethics in tourism, the role of tourism in ideological production, consumer culture, and matters of tourism epistemology and pedagogy.

Michael Clancy is Professor and Chair of the Department of Politics and Government at the University of Hartford, Connecticut, USA, where he also directs the International Studies Program. His research focuses on the international political economy of tourism. He has written on tourism and national development in Mexico and the Republic of Ireland, tourism and nation branding, and tourism and global commodity chains.

Sarah Quinlan Cutler is a PhD candidate in Geography and Environmental Studies, Wilfrid Laurier University, Waterloo, Ontario, Canada. Her research interests include responsible tourism, tourism experiences, experience research methods and Peruvian tourism.

Cynthia Shirley Deale is Professor in the School of Hospitality Leadership at East Carolina University, USA. Her research interests include the scholarship of teaching and learning, sustainability, and tourism and hospitality management practices. She serves as a facilitator for the Certified Hospitality Educator workshop for the American Hotel and Lodging Association, and has served in a leadership capacity on the board of the International Council of Hotel, Restaurant, and Institutional Education.

Tareq Emtairah is Assistant Professor at the IIIEE/Lund University, Sweden. His work includes education and research related to cleaner technologies and the integration of sustainability into sectoral strategies for transport, housing and tourism. He holds a PhD in Environmental Management and Policy from Lund University in Sweden. Currently he is the head of the Regional Center for Renewable Energy and Energy Efficiency (RCREEE) in Cairo.

Wouter Geerts is a PhD candidate at Royal Holloway, University of London. His research interests lie in the fields of sustainability, hospitality and tourism/travel, with a particular interest in environmental behaviour change. For his PhD he is conducting interdisciplinary research to investigate the impact of travelling for work on individuals' pro-environmental behaviour. He has presented papers at international conferences in the disciplines of geography and management, and co-organized the Royal Geographical Society's PhD colloquium on tourism research at the University of Surrey in 2012.

C. Michael Hall is Professor in the Department of Management, Marketing and Entrepreneurship, University of Canterbury, New Zealand; Docent, Department of Geography, University of Oulu, Finland; Visiting Professor, Linneaus University, Kalmar, Sweden; and a Senior Research Fellow, School of Tourism and Hospitality, University of Johannesburg, South Africa. Co-editor of *Current Issues in Tourism*, he has long standing interests in tourism, global environmental change and gastronomy.

Andrew Holden is Professor of Environment and Tourism and Director of the Institute for Tourism Research (INTOUR) at Bedfordshire University in England. His research focuses on the interaction between human behaviour and the natural environment in the context of tourism. He has worked in various countries on tourism development projects and is the author of several texts, including *Environment and Tourism*, and the *Routledge Handbook of Tourism and the Environment*.

Carol Kline is Associate Professor and currently works at East Carolina University, USA. She received her PhD in Parks, Recreation and Tourism Management from North Carolina State University, where she taught classes and worked for Tourism Extension. Her research interests focus on rural tourism development including how to create a supportive environment for tourism entrepreneurs, the role of sustainable food systems in tourism, and early tourist markets in burgeoning and rural destinations.

Whitney Knollenberg is a PhD candidate in the Department of Hospitality and Tourism Management at Virginia Tech, USA. She earned her MS in Sustainable Tourism at East Carolina University and her BSc in Parks, Recreation, and Tourism Resources at Michigan State University. Her research interests include the connection between tourism and sustainable food systems, tourism planning, and the role of policy, power and partnerships in tourism development.

Maria Koleth is a PhD candidate in the Department of Gender and Cultural studies at the University of Sydney, Australia. Her current research focuses on popular practices of development, such as volunteer tourism, and their implications for development ethics.

Brent Lovelock is Associate Professor in the Department of Tourism at the University of Otago, New Zealand. His research considers sustainable tourism from a broad perspective, including ethical, social, political and ecological aspects.

Kirsten Lovelock is a medical anthropologist and Senior Research Fellow in the Department of Preventive and Social Medicine at the University of Otago, New Zealand. She has research interests in: environmental health, occupational health, social inequality and health systems, and healthcare pathways and ethics in relation to these fields. Brent and Kirsten recently published a co-authored book: *The Ethics of Tourism: Critical and Applied Perspectives* (Routledge), and are currently conducting research focusing on medical tourism in New Zealand and India.

Sheila Malone is a Lecturer in Marketing at Lancaster University Management School, UK. Her research interests lie in the area of consumer behaviour and tourism experiences. Specifically she focuses on ethical tourism, the role of emotion in tourists' ethical decision-making processes and the interplay between hedonism and ethical tourism practices. Her recent research interests have focused on ethical luxury services in a tourism context.

Vania Ranjbar has been involved with HIV/AIDS work and charities in South Africa since 2003. Her research generally combines social and health psychology, most commonly employing qualitative methods. Her PhD research, a discourse analytic ethnography, investigated AIDS-related bereavement among aid workers in South Africa. After a decade working in an international context she returned to Sweden in 2012 and currently works as a researcher for the Unit of Social Medicine, University of Gothenburg, Sweden, and Angered Local Hospital.

Clare Weeden is Senior Lecturer in Tourism at the University of Brighton, UK. Her PhD explored the influence of personal values on the holiday choice decisions of ethical consumers. The majority of her research focuses on ethical and responsible tourism, the motivations of responsible tourists, corporate social responsibility in tourism, and marketing and promotion of SME-responsible tour operators. Her most recent publication is a research monograph entitled *Responsible Tourist Behaviour*, published by Routledge in August 2013.

Foreword

The current place of ethics in the tourism studies order of things may be characterised as marginal at best. It represents what surely must be a very small percentage of the gamut of tourism scholarship. On one hand, the lack of sustained work in this area is difficult to fathom given the frequency and depth of moral issues in tourism. Sex tourism, theft, slum tourism, corruption, greed and so on, are all persistent themes in tourism that will only get worse as the industry continues to grow. On the other hand, it is not such a stretch to see why there is such a dearth of work in this area, as amply explained in the Introduction of this volume. Weeden and Boluk write, 'interest about ethical issues in wider society comes to an almost-abrupt halt when applied to holidays and consumers' leisure time and activities'. Several years of personal observation in this area point to the fact that practitioners will embrace ethics if it provides competitive advantage; tourists actively pursue what is in their own best interests; and tourism theorists have other priorities, tied to the more traditional disciplines such as anthropology, geography, sociology and business.

The reasons for a slower-than-expected flurry of interest in tourism ethics research are not altogether clear. Part of the problem is that tourism theorists often place ethics into some other category like responsible tourism, or think about what is right and wrong in tourism according to various certification schemes. The argument for these approaches is that we ought to be more concerned about the 'doing' of tourism. However, we should be careful about this for a couple of reasons. First, a focus solely on responsibility takes us away from the history and richness that ethics provides in explaining the subtleties of human nature. Failing to utilise this rich tradition serves only to flatten and narrow our options. Second, and related to the first, is that there are preconditions for being responsible. One of these is that responsibility can only be achieved by first working from a moral framework. You really cannot have one (responsible) without the other (ethics).

Another part of the problem is that the field does not offer a home for many of the more specialised topics of tourism, like tourism ethics – not that such is warranted at this time. But papers scattered among an array of different journals makes it more difficult to build consistency in any area. And even if specialised areas do have a home (e.g. *Journal of Ecotourism*), the vast majority of papers are submitted to other more general tourism periodicals. Still another issue stems from the fact that very few tourism scholars have been formally schooled in moral philosophy. It is natural for scholars to steer clear of ethics without the confidence to venture forward, while at the same time staying true to the disciplines (as above) in which they have been taught. What will continue to hold ethics back in tourism is the fact that few schools are presently turning out PhDs in this area, and this is a trend not likely to change soon.

What is even more unsettling is that tourism impacts – the very backbone of the tourism studies field – are all, in fact, ethical issues. Yet we continue to marginalise ethics as something too impractical to solve tourism's real-world issues. We simply need to take notice of other disciplines like medicine and healthcare, business, environment, marketing, law and engineering, where ethics holds a central position both in theory and practice. But the main or primary focus of these other fields is not the deliberate pursuit of pleasure, like it is in tourism. Ethics in tourism may be uncomfortable and conveniently marginalised because it simply gets in the way.

Taking the discussion on other disciplines one step further, I have long felt that tourism studies will not advance unless we are willing to embrace the most essential theories of the disciplines that help define who we are as a field. I continue to argue (Fennell 2006b) that moral philosophy is one of these other important fields of enquiry – like geography, sociology, anthropology, business and so on – that is essential to getting to the root of the many problems that plague tourism. What is unique about ethics is that more than any of these other fields of enquiry, it gets to the job of illuminating the tensions that exist between what is good and bad, right and wrong, and authentic and inauthentic in tourism. It does this by forcing us to examine the human condition first, knowing something of ourselves, as a way to understanding tourism (see Przclawski 1996; and Wheeller 2004).

In tourism, the freedom to move in space and time also comes packaged with the freedom to act in ways that may or may not be consistent with the ways we act at home. At home, the shadow of the future (Axelrod 1984), or the ongoing interaction with the same people over and over again, compels us to treat them in a manner that we ourselves would like to be treated. When we travel, taking this cloak of morality with us gives us the opportunity to hone or even advance our moral practices in recognition of the fact that we are temporary visitors in places that we should honour and protect. We demonstrate cooperation with other people and commitment to moral principles. By contrast, we can choose to leave the moral cloak at home and adopt a type of behaviour where the focus is squarely on our own needs at the expense of other people and other things. In the vernacular of reciprocal altruism, this means that we are prepared to cheat others (local people or tourism brokers), or vice versa, in the pursuit of our own self-interested projects (Fennell 2006a).

Both ends of the continuum above, the ethical-by-choice and the not-so-ethical-by-choice, are in need of investigation on many different levels. This rather wide spectrum makes it absolutely essential that we continue to educate theorists, practitioners and, of course, tourists, about what is good, right and authentic in the context of tourism. In this vein, the present volume is well positioned to move the debate forward. The chapters, as constituent parts, demonstrate the depth and range of moral issues that are open for consideration in tourism. The book as a whole is important because it keeps ethics timely, while at the same time forcing us to ask bigger and better questions about how to move the tourism ethics agenda forward.

David A. Fennell
Professor, Tourism and Environment
Brock University, Canada

References

Axelrod, R. (1984) *The Evolution of Cooperation*. New York: Basic Books.

Fennell, D.A. (2006a) 'Evolution in Tourism: The Theory of Reciprocal Altruism and Tourist–Host Interactions' in *Current Issues in Tourism*, 9 (2): 105–24.

Fennell, D.A. (2006b) *Tourism Ethics*. Clevedon: Channel View Publications.

Przeclawski, K. (1996) 'Deontology of Tourism' in *Progress in Tourism and Hospitality Research*, 2: 239–45.

Wheeller, B. (2004) 'The Truth? The Hole Truth. Everything But The Truth. Tourism and Knowledge: A Septic Sceptic's Perspective' in *Current Issues in Tourism*, 7 (6): 467–77.

Acknowledgements

The editors wish to thank a number of people who have offered their guidance and patronage from the outset to make this volume possible. First, we would like to sincerely thank all of our chapter contributors, who have become good friends and have made this an enjoyable and fruitful experience. This volume would not have been possible without your contributions and expertise. Thanks to Alison Gill for inviting us to write a Call for Papers for the Association of American Geographers Conference in New York 2012. Many thanks to Michael Hall and Tim Coles for inviting us to be part of their exciting Routledge Critical Studies in Tourism, Business and Management. We would also like to thank Micha Lück for his time and feedback upon review of our initial book proposal. A special thanks to Editor, Emma Travis, and Editorial Assistant, Pippa Mullins, at Taylor and Francis for their kind support and patience over the last few months. A special thank you also goes to David Fennell for leading the way in tourism ethics research. We are very grateful that David took the time to compose a foreword that powerfully articulates the significance of research in the area of ethics in tourism, and succinctly *synthesises* the contents of this volume. Finally, thanks to all of the blind reviewers who generously gave of their time and expertise.

Karla Boluk

I would like to express my thanks to Dalarna University, Sweden, who supported me during the initial part of this book's development and who funded my conference attendance to the AAG in 2012. My personal gratitude goes to my parents Catherine and David Boluk for their love and support, siblings Jennifer and Lucas, and husband Gary. This book is dedicated to those who contemplate their impacts in the context of tourism and consider ethics in their decision-making, as well as researchers who take on the challenge of exploring this intricate domain in an effort to discover how tourism can be a force for sustainable change.

Clare Weeden

Thanks go to the School of Sport and Service Management, University of

Brighton, UK, for supporting the development of this edited book through the American Association of Geographers Annual Conference in New York, February 2012. Special thanks go to the willing reviewers of the chapters and, of course, to our wonderful and generous contributors.

1 Introduction

Managing ethical consumption in tourism – compromises and tensions

Clare Weeden and Karla Boluk

Discussions about the negative and widespread impacts of tourism far outweigh those that reflect the various opportunities created for people, culture and environment in the tourism process, and thus provide a rationale for debate about ethics in tourism. Specifically, many commentators have noted the negative consequences associated with the development, management and operational activities of the tourism industry (Ashley *et al.* 2001; Mowforth and Munt 2003; Higgins-Desbiolles 2010). Other research considers tourism to be highly exploitative (Sánchez Taylor 2010), socially damaging (Tucker and Akama 2009) and notably a 'justice issue' (D'Sa 1999: 68). Fewer studies have explored the opportunities associated with tourism development. For example, deliberations in the areas of peace (Ap and Var 1990; Moufakkir and Kelly 2010), poverty alleviation (Spenceley and Goodwin 2007; Muganda *et al.* 2010; Boluk 2011a; Scheyvens and Russell 2012), transformational tourism (Reisinger 2013) and opportunities to facilitate meaningful and co-transformational interactions (Pritchard *et al.* 2011), while not entirely overlooked by researchers, have been largely ignored.

In response to existing criticism regarding tourism development, the literature charts a plethora of mitigative approaches, usually presented under the umbrella term 'alternative tourism'. Such approaches demonstrate (at least on the surface) the simultaneous efforts made by tourism producers and tourism consumers towards sustainability goals. While these alternative ways of conducting and consuming tourism have their critics (see for example, Wheeller 1991; Butcher 2003), they are arguably reflective of what Hartwick (2000: 1179) refers to as an 'ethical turn' in society, which increasingly challenges the dominant postmodernist support for mass consumption.

Despite these alternative approaches it appears as though the tourism industry, producers, and tourism consumers, have not responded convincingly to calls for more responsibility in tourism. Correspondingly, Fennell (2006; 2008) argues that ethics in tourism has not been given adequate research attention and thus clearly has the potential for greater investigation. Indeed, when ethics is discussed in a tourism context it often focuses on topical and obvious issues such as sex tourism (see Opperman 1998; Kibicho 2009), the impact of tourism development on the environment (Kavallinis and Pizam 1994; Briassoulis 2000; Butler

2000; Mbaiwa 2003; Amuquandoh 2010), or latterly, debates regarding tourism's role in climate change. Occasionally, discussions address the social aspects of tourism but these tend to centre on tourist conduct, such as 'inappropriate dress or behaviour, [the] demonstration effect, and cross-cultural conflict or cultural erosion' (Scheyvens and Momsen 2008: 31).

In contrast, a great deal of deliberation has centred on ethics in a variety of disciplines outside the tourism field. In fact, nearly a decade ago, Popke (2006: 504) noted the sheer volume of academic journal articles (in the order of 1,500 annually), suggesting that ethical conversations are not just popular in academia but are also taking place in the social domain, and discussed by 'bankruptcy lawyers, money managers, judges and dentists, and applied to our sporting events, our militaries, and even our space agencies'. However, interest in ethical issues in wider society comes to an almost-abrupt halt when applied to holidays and consumers' leisure time and activities. Indeed, it appears there is little interest from tourists in being responsible on holiday, as indicated in a survey by the Association of British Travel Agents (ABTA), which revealed 41 per cent of UK consumers preferred to relax on holiday instead of thinking about green issues, and 48 per cent of them had no opinion as to whether or not holidays should have a social or environmental rating (ABTA 2010).

Interestingly, titles and associations have appeared to be significant to some tourism consumers, resulting in the removal of the term 'tourist' from anything affiliated with the concept of holidaying. Consequently, tourists have been redefined as adventurers, fieldwork assistants, exploraholics, volunteers and travellers, all of whom Birkett (2002: 3) argues are 'the same people' *doing* 'the same thing'. Yet, such disassociations have the potential to eliminate perceived guilt felt by consumers. The disassociation from mass tourism is also important to some travellers (McCabe 2005), giving rise to distinctive forms of tourism such as backpacker tourism (O'Reilly 2006). Whether this reflects a heightened awareness of the negative consequences associated with (mass) tourism, or an interest in demonstrating that one is 'ethical', remains questionable.

The lack of research in the area of ethics and tourism has provided the impetus for this volume. As noted by David Fennell in the Foreword, it is apparent that tourism scholars have avoided engagement with ethical issues in tourism, arguably demonstrating a lack of confidence. Specifically, there is a dearth of research exploring the compromise and tension experienced by consumers in their tourism and travel decision-making. This focus framed a call for papers, submitted by the editors, for the Association of American Geographers (AAG) Annual Meeting held in February 2012, in New York, USA. This call generated interest from scholars around the world and hence facilitated an interactive working session at the AAG, where contributors were challenged to present their papers in five minutes, with an hour of follow-on discussion. The high level of interest and enthusiasm from these scholars led directly to this edited book, and the high quality of the contributing chapters is testament to the energy and commitment of all the people involved.

To many people, 'ethical debate in tourism' is something of an oxymoron. The former addresses issues of right and wrong, or 'inquiry into the nature and grounds of morality, where morality means moral judgements, standards, and rules and conduct' (Tsalikis and Fritzsche 1989: 696). In contrast, the latter involves mostly selfish desires centring on fun, relaxation and 'getting away from it all'. Ethical consumption in tourism, often implicit within holidays sold under the banner of eco, responsible, Pro-Poor or Fair Trade, *is often unwelcome* because it places additional pressure on travellers to make appropriate decisions at a time when hedonic motivations are often prioritised and supersede all others.

Of additional challenge, especially to those supporting the drive for ethical holidays, is the relatively limited availability, and sometimes expensive nature of responsible tourism 'products', as well as the extremely complex nature of the various component parts of the holiday experience which consumers find challenging to navigate. For instance, transport, accommodation and attractions often make great claims for sustainable practices, but these are difficult for people to identify, understand or believe (Weeden 2013). Such issues are at the heart of this book, which aims to provide a rich interpretation of the complexity of ethical consumption in tourism, insight into discourses influencing travellers and their decision-making, and offers deep insight into the key challenges facing stakeholders keen to support the production and consumption of responsible tourism.

Managing ethical consumption in tourism

Before introducing the contributions in this edited book it is important to provide a closer examination of the debates and discussions involved in managing ethical consumption in tourism. As noted earlier, the benefits and consequences linked to tourism evoke strong emotion. For instance, despite its enriching benefits, in areas such as infrastructure and service improvement, employment opportunity and wealth creation (Andereck *et al.* 2005), tourism is acknowledged as potentially exploitative (Sánchez Taylor 2010), and capable of perpetuating inequity and exclusion (Carlisle 2010; Cole and Morgan 2010; Hall 2010). In addition, while tourism is positively embraced by a range of community stakeholders (Nunkoo and Ramkissoon 2010), it is the damaging effects associated with mass tourism that tend to dominate tourism inquiry.[1] For instance, resident attitudes towards increased traffic, crowds, congestion and urban sprawl in towns and cities are popular foci of tourism research in developed countries (see for example, Andereck *et al.* 2005; Choi and Murray 2010), but its potential to irreversibly damage pristine landscapes around the globe is the most often recalled characteristic of tourism's environmental impact. In particular, uncontrolled visitor numbers in national parks, unregulated tourism development around sensitive ecosystems, and diversion of water from local food production for tourists' leisure use (Kline *et al.* this volume), raise familiar questions over habitat loss, disturbance to animals, and issues of environmental justice (Sasidharan *et al.* 2002; Amuquandoh 2010; Cole 2013; Holden and Fennell 2013).

Not only is the environment impacted by the production and consumption of tourism. There are also human costs, and many commentators question tourism's ability to manage itself ethically, especially when it concerns the impact on local communities (Tosun 2002; Andereck et al. 2005), the appropriation of culture and heritage for the purpose of tourism (Kirtsoglou and Theodossopoulos 2004), and the consequences for community cohesion of poor tourism planning and ineffective business practice (Honey and Gilpin 2009). Further debates examine the ethics of representation and commodification of indigenous peoples (Caton, this volume; Saarinen and Niskala 2009), disempowerment of local communities through international investment (Mbaiwa 2003), and prevention of access to farmland and other critical resources as a consequence of conserving wildlife for tourism (Rutten 2002; Snyder and Sulle 2011).

While not claiming to be an exhaustive list of the ethical dilemmas associated with travel and tourism, they provide some explanation as to why tourism is often accused of exacerbating global disparities of wealth and reinforcing structural inequality (Britton 1982; O'Hare and Barrett 1999; Chong 2005; Scheyvens 2007; Brown and Hall 2008). Since there is virtually nowhere on the planet that has not felt the 'effects of modernity through tourism development' (Smith and Duffy, 2003: 2), such inequities are problematic for a range of stakeholders, many of whom have called for a more 'just' and equitable form of tourism (de Kadt 1979; Krippendorf 1987; Hultsman 1995). In response to such calls, a plethora of 'alternatives' have been offered to the market place. These include 'sustainable tourism' (Eagles et al. 2002), 'eco-tourism' (Fennell 1999), 'volunteer tourism' (Wearing 2001, 2004), 'ethical tourism' (Tribe 2002; Weeden 2002; Butcher 2003; Fennell 2006), 'responsible tourism' (Wheeller 1991; Wright 2006), 'pro-poor tourism' (Ashley et al. 2001; Scheyvens 1999, 2002, 2007), and 'fair trade tourism' (Kalisch 2001, 2010; Boluk 2011a, 2011b).

While many of these represent a genuine desire to develop and support a tourism that offers beneficial and equitable exchange, critics continue to question their relevance, utility and effectiveness (see Wheeller 1991; Butcher 2003). For example, some commentators dismiss them for being ineffectual solutions to the 'problem' of mass tourism, and symptomatic of a New Puritan movement, where an application of ethics to tourism is (cynically) couched in the discourse of virtue and the promise of doing good (Butcher 2003; Lisle 2008). As Lisle notes (2008: 5), 'not only does ethical tourism solve all the problems caused by mass tourism, it also makes you a better person and the world a better place'. Such debates are not without support (see Mowforth and Munt 2009). However, there is also evidence that an increasing number of stakeholders are acknowledging the need for an ethical approach to travel and tourism. For example, while still a minority in terms of holiday sales, responsible tourism products exist in the portfolio of many multinational tour operators. Similarly, the social and environmental impacts of tourism are becoming more prominent in the minds of tourists, with 47 per cent of UK travellers believing their holidays should directly benefit local people and economies (ABTA 2010).

Ethical turn in society

Concerns over social and trade justice, and debates over responsibility in tourism, have their foundation in wider societal debates. For example, ethical consumption is conceptualised as a set of practices where global solidarity can be (voluntarily or routinely) displayed and enacted, and 'through which unequal power relations are constituted, reproduced and contested' (Barnett *et al.* 2005: 41). Such interpretations rest upon the knowledge that individuals across a globalised world are interconnected through daily consumption habits (Hartwick 2000), and a renewed conception of spatial connections has prompted further debate regarding the ethical issues involved in production, supply and purchase of goods and services (Shaw *et al.* 2000; Szmigin *et al.* 2009; Shaw and Riach 2011).

Consumers in the affluent north have long demonstrated an interest in ethical consumption (e.g. Seif 2001; Goodwin and Francis 2003; Harrison *et al.* 2005; Nicholls and Opal 2005). For example, anxiety over the sustainability of the environment (Carson 1962; Wilkins 1999) has led to increased production of organic foods and ecologically benign goods, an emphasis on buying locally (e.g. within a 30-mile radius), and greater participation in recycling programmes. Further to environmental priorities, people exhibit increasing concern for the treatment of human beings (Peattie 1992) in the Majority World, typically when social issues are addressed in the media. Such attention has also prompted consumer participation in anti-sweat shop campaigns, have persuaded consumers to boycott specific companies and/or support (buycott) companies that are transparent in their business operations.

The development of consumer interest in environmental (Holden, this volume), and human and non-human animal rights in the production and supply of commodities and services has also influenced consumer choice in tourism, typically in discussions of fair trade in tourism (Kalisch 2013), ethical responsibilities of tour operators (Budeanu 2005), or moral selving of volunteer tourists (Boluk and Ranjbar, this volume). Further evidence exists in consumer willingness to boycott holiday organisations, which encouraged Tourism Concern, a UK-based non-governmental organisation (NGO), to encourage a boycott against travelling to Burma in protest against the repressive military junta. Similarly, consumer demand has encouraged a growing number of commercial operators in the UK, Europe, Australia and the USA to introduce sustainability checklists for their supply chains, and growing numbers of consumers are considering the responsible reputation of travel companies in their decision to buy 'guilt free' holidays.

Thematic structure of the book

This book addresses ethical consumption in tourism from three different, yet interconnected, perspectives, with the intention that readers will develop a holistic appreciation of the complexity associated with contemporary debates. It is organised as follows: Part I, 'debates on ethical consumption in tourism'; Part II, 'situating the self in ethical consumption'; and Part III, 'helping consumers make ethical decisions'.

Part I presents a set of conceptual pieces that question popular interpretations of 'ethics' in tourism. Building on these deliberations, Part II offers a set of empirical studies into why, and also how, individuals construct ethical self-identities in tourism. Finally, Part III offers acumen and practical insight into the many challenges associated with helping tourists make responsible and potentially more ethical choices on holiday.

A key differential of this book is the presentation of an original activity, with the inclusion of two or three discussion questions at the end of each chapter. This innovation is specifically designed to assist instructors in the facilitation of group or classroom discussions on the theory and praxis of ethical consumption in tourism. Each activity offers a short *précis* in the form of either a case study, scenario or discussion point, with questions and prompts for consideration. These provide lecturers and associates with a class-based tool kit to enable comprehensive curriculum delivery. Additionally, most chapters provide two or three recommended readings and/or websites for further discussion.

Research contained in the following 14 chapters is produced by scholars from around the world who are interested in understanding ethics in tourism from a theoretical, philosophical, conceptual and applied perspective. A variety of international contexts are explored, in both the developed and developing world, including Uganda (Chapter 5), Sweden (Chapter 6), the USA (Chapter 7), Australian visitors to Cambodia and Peru (Chapter 8), South Africa (Chapter 9) and the UK (Chapter 12). In addition to the applied nature of several chapters, which make the book useful to a range of corporate, NGO and third sector stakeholders, the multidisciplinary nature of the research offers insight to scholars across a broad range of subjects, including tourism management, geography, consumer behaviour, sustainability and ethics, cultural studies, sociology and development studies. The following summaries are intended to introduce each chapter and highlight some of the key contributions made.

Part I: debates on ethical consumption in tourism

This first part contains four chapters that examine the contemporary discussions underpinning ethical consumption in tourism from a philosophical and conceptual perspective. Specifically, the chapters theorise the politics of consumption and critique the relationship between ethics and philosophy, environmentalism and aesthetics, and explore the tension, compromise and perhaps contradictions apparent in the decision-making process for all stakeholders involved in sustainable and ethical tourism consumption.

The first chapter in this part, written by Kellee Caton, reflects on historic connections between Western preoccupation with material consumerism, ethical identity construction, and tourism. She reviews the binary dilemma of individu-

Introduction: managing ethical consumption 7

als who struggle with what they conceive to be a moral dilemma – having fun on holiday while avoiding the potentially unjust and exploitative consequences of such activities. For Caton, the impacts cannot be solved through transactional ethics, and she concludes the chapter by calling for a greater emphasis on philosophical solutions to help people resolve any tension between 'nurturing self and doing right by others'.

The following chapter, written by Michael Hall, provides a comprehensive review of ethical consumption and its interpretation within tourism and hospitality discourse. Contextualised by the imperative of climate change and the impact of tourism's continued and inexorable growth, the chapter critiques consumer sovereignty and the failure of neo-classical economic models to manage consumption at sustainable levels. Using the fair trade movement, Hall explores the consumerist appeal of commodity fetishism, debates corporate exploitation of the links between capitalism and conservation, and discusses the importance of promoting positive and beneficial behaviour change through a combination of social marketing and public policy nudging.

The third chapter in this part is written by Michael Clancy and examines the evolution and development of slow tourism in a world where rationalisation, standardisation and homogeneity are the norm. With a detailed review of critical debates in slow travel literature, Clancy discusses the concept and practice of slow tourism, identifies key typologies, and acknowledges connections between holidaying slowly, increased promotion of responsible tourism, and wider practices of ethical consumption. In its reflection on the symbolic value of slow travel, the chapter considers Veblen's conspicuous consumption and Hirsch's concept of positional goods, and considers whether exclusivity or moral virtue are relevant to ethical choice in tourism. In conclusion, the chapter questions the economic sustainability of slow tourism and considers whether slow travel is as beneficial for host communities and providers of slow tourism experiences as they are for tourists.

Andrew Holden writes the final chapter in this first part. The chapter begins with a reference to UNEP's (2011) report on tourism's importance for a Green Economy and considers the characteristics of tourism demand that make such an objective uniquely challenging. Holden critiques the tension between a growing global dependency on tourism to alleviate poverty and the increasing demand for leisure experiences that damage the environment. The chapter continues with a reflection on the spatial and non-spatial elements of community and global interconnectivity, and uses the work of Thoreau, Muir and Leopold to chart the development of an environmental ethic and its influence on our appreciation and use of nature for tourism. After a comprehensive review of the ethical perspectives of egoism, utilitarianism and altruism, Holden discusses the principles of deep ecology (Naess 1973) and the conservation ethic, and argues for a re-evaluation of how tourism appropriates the natural environment for human benefit. The chapter concludes with a case study from Uganda, which illustrates the ongoing debate over land use values and the intrinsic right of nature to an existence independent of human value.

Part II: situating the self in ethical consumption

The second part of this book builds on the discussions presented in Part I, and presents cutting edge research on the different ways responsible travellers attempt to navigate the marketplace. These chapters offer specific examples of the trade-offs and self-negotiations that take place before, during and after holiday decision-making. The five chapters in this part comprise both theoretical and empirical investigations into the complexities of individual action and identity construction through tourism. The former is examined through tourists' destination behaviour, while the latter is explored through the notion of ethical selving in the context of volunteering, and the role of emotion in ethical travel choices.

The first chapter in this part is written by Adriana Budeanu and Tareq Emtairah, and reports on an empirical study of the purchasing choices of tourists visiting Sweden. After a detailed overview of the debates and ambiguities revealed by studies of tourists' environmental awareness and sustainable behaviour, Budeanu argues for greater research attention on the destination context of tourist choices and for parallel streams of consumption such as transport, accommodation and leisure activities to be included in such inquiry. In her view, such knowledge is essential to an accurate evaluation of tourists' environmental behaviour, which in turn provides a more effective outcome for sustainable tourism policy and practice. Of most significance, Budeanu introduces the term *environmentally preferred option (EPO)*, to situate tourist choices across a range of services in holiday destinations. Such nuanced information directly informs consumers and also encourages public and private tourist organisations to make explicit use of such knowledge. The chapter concludes with a discussion of the methodological implications of the findings, which make a significant contribution to discourse on sustainable tourist consumption.

Continuing with the theme of sustainable action in tourism, the following chapter, written by Carol Kline, Whitney Kollenberg and Cynthia Deale, argues that ethical food providers need to work collaboratively with tourism suppliers to provide a comprehensive set of sustainable food tourism experiences. Using the framework of stakeholder theory, the chapter reflects on a societal trend to 'eat locally', citing an increase in farmers' markets, farm-to-fork restaurants, and a consideration of food miles as evidence of increasing consumer interest in sustainable food systems. The authors note the current lack of synergy between tourism and sustainable food supply chains and argue many tourists seek to make ethical decisions about food and eating on holiday, and enjoy observing or taking part in harvesting and food processing activities, but may be ignorant as to how to access such experiences while away from home. The chapter makes reference to fundamental connections between ethical issues in tourism and food production, such as low wage cultures, propensities to damage the environment, and persistent consumer demand for discounted goods and services, and the subsequent impact on production practices and animal welfare standards. Kline *et al.* deliver a comprehensive overview of research into sustainable options in the

Introduction: managing ethical consumption 9

food value chain and conclude with recommendations for future research directions to encourage and support ethical eating decisions in tourism.

Focusing on volunteer tourists' pursuit of an ethical (self) identity in tourism, Maria Koleth's chapter unpicks the complex motivations of development tourists. Situating her thesis within the commodification of development and a neo-liberal economy, and connections with moral selving, Koleth presents an empirical study of Australian volunteer tourists to Cambodia and Peru. Using Foucault's (2010) self-as-enterprise as a framework, the chapter presents volunteer tourists' desire to competitively differentiate themselves in the market, and considers the ethics of a neo-liberal agenda whose goal is the accumulation of human capital rather than eradication of global poverty. Koleth critiques development tourism and questions the motivations of volunteering organisations responsible for creating and encouraging demand for such experiences, while at the same time acknowledging the often genuinely altruistic motivations of the volunteers. The chapter concludes with recommendations for future research.

Continuing the theme of self-as-enterprise, the empirical contribution in the chapter by Karla Boluk and Vania Ranjbar addresses the motivations of volunteers to Cotlands, a non-governmental organisation (NGO) in South Africa that cares for vulnerable children and families affected by HIV/AIDS in residential and community outreach programmes. Employing a discourse analysis, the chapter critiques ten volunteers' motivations against four emergent themes: personal fulfilment; detachment and reflections on returning home; distancing discourse; and preferred ways to contribute. Highlighting the unique contribution of the study, Boluk and Ranjbar reveal volunteers are motivated less by altruism than by actions to construct an ethical self-construct.

The final chapter in this part is written by Sheila Malone and continues the theme of ethical identity construction, but from a social psychology perspective, with the intention to contribute to debate on the knowledge gap with regard to ethical intention and behaviour. Focusing on the motivational role of positive and negative emotion states in tourists' ethical choices, the chapter presents a critical review of the connections between ethical tourism experiences and the concept of emotion in consumers' ethical choice formation. The chapter concludes with a reflection on the significance of understanding emotional attachment for tourism organisations that seek to create, sustain and strengthen customer loyalty and commitment to ethical consumption practices. Recommendations for future research offer further useful insight.

Part III: helping consumers make ethical decisions

The third and final part of this edited book addresses what we believe to be a significant gap in the ethical tourism literature – the need to offer practical help and support to consumers in their responsible travel choices. As such, these chapters present discussions on the role of ethical information in travel guidebooks, the challenge of persuading business travellers to be environmentally aware when travelling for work, and the ethical consequences of medical tourism. The final chapter examines how marketing and communications can be harnessed to help consumers make ethical decisions.

The first chapter in this part, written by Sarah Quinlan Cutler, offers a valuable contribution to knowledge on the potential for travel guidebooks to inform and encourage tourists to engage in ethical travel behaviour. Noting tourists' search for information as a crucial element in decision-making, Cutler argues that although guidebooks influence perceptions of destinations and direct the travel practices of millions of tourists, little is known about the range and depth of available ethical information in guidebooks or how it is communicated. The chapter addresses this gap through the presentation of a study of ethical tourism content in the text of commercial guidebooks about Peru. Using content analysis, Cutler provides a detailed and critical discussion of the ethical information, and highlights changes in content between the 2004 and 2012 editions. The discussion queries who is responsible for ethical information in guidebooks, and questions to what extent tourists respond positively to such material. The chapter concludes with a thoughtful critique of the study and offers recommendations for further research.

The following chapter, written by Wouter Geerts, examines an area of research significantly neglected in sustainable tourism literature. The chapter presents an empirical study of business travellers to hotels in London, UK, and assesses these individuals' attitudes to travelling and staying in hotels, and the extent to which their hypermobility and work responsibilities impact upon pro-environmental choice. The chapter provides a useful critique of business travel, identifies factors important to business travellers' decision-making, and considers how these differ from the priorities of leisure tourists. It also offers insight into business travellers' willingness to engage in environmental practices when away from home. Geerts' study makes a significant contribution to knowledge about business travellers' environmental considerations, information seen as essential in the encouragement and support of demand and provision of environmentally efficient hotels.

The third chapter in this part is written by Kirsten Lovelock and Brent Lovelock and addresses an equally under-researched area of study, but one currently attracting increased attention because of its inherent ethical impact: medical tourism. The authors present a comprehensive review of medical tourism's origins and development, and discuss the ethical consequences for both provider and departure countries of a neo-liberal agenda that commodifies healthcare. Using content analysis, Lovelock and Lovelock evaluate internet blogs (written by tourists about their experiences of non-essential medical treatment overseas) to address fundamental questions on issues such as power, access and affordability of healthcare at home and abroad, responsibility and social justice.

The first three chapters in this part of the book have reported on research and information that can be used by a range of stakeholders to encourage more tourists to take ethical decisions and more organisations to help satisfy this demand. This final chapter, written by Clare Weeden, takes the discussion to the logical next step – how can marketing, a key tool in influencing human behaviour, be effectively harnessed in the achievement of these objectives? This chapter reviews the ethical dilemmas commonly associated with tourism and discusses how the attitude–behaviour gap poses a significant challenge to those charged with marketing responsible tourism. The chapter questions whether public scepticism and cynicism over marketing's reputation for unethical practice can be overcome to enable a more effective communication of the benefits of adopting ethical consumption in tourism.

As established at the beginning of this introduction, our aim with this volume is to contribute to existing dialogue on ethical consumption in tourism and to explore emerging debates on how the (ethical) self is recognised and in what way consumers make responsible travel decisions. We hope readers use this text to help them make up their own mind about what it means to be ethical in tourism, and offer the insights contained in this book as an aid to this objective.

Note

1 For critiques of mass tourism being 'wrong' or bad, see Clarke (1997), Fennell (2006), and Diamantis and Ladkin (1999).

References

ABTA (2010) 'The ABTA 2010 Consumer Survey: Insights into the Modern Consumer', online, available at: www.abta.com/news-and-views/press-zone/sustainability-an-essential-not-an-optional-extra (accessed 12 October 2012).

Amuquandoh, F.E. (2010) 'Residents' Perception of the Environmental Impacts of Tourism in the Lake Bosomtwe Basin, Ghana' in *Journal of Sustainable Tourism*, 18 (2): 223–38.

Andereck, K.L., Valentine, K.M., Knopf, R.C. and Vogt, C.A. (2005) 'Residents' Perception of Community Tourism's Impact' in *Annals of Tourism Research*, 32 (4): 1056–76.

Ap, J. and Var, T. (1990) 'Does Tourism Promote World Peace?' in *Tourism Management*, 11 (3): 267–73.

Ashley, C., Roe, D. and Goodwin, H. (2001) 'Pro-poor Tourism Strategies: Making Tourism Work for the Poor. A Review of Experience', Pro-poor Tourism report No. 1, Overseas Development Institute for Environment and Development, London, and Centre for Responsible Tourism, University of Greenwich. Online, available at: www.propoortourism.org.uk/ppt_report.pdf (accessed 2 January 2013).

Barnett, C., Cloke, P., Clarke, N. and Malpass, A. (2005) 'Consuming Ethics: Articulating the Subjects and Spaces of Ethical Consumption' in *Antipode*, 37 (1): 23–45.

Birkett, D. (2002) 'Re-branding the Tourist' in T. Jenkins (ed.) *Ethical Tourism: Who Benefits? (Debating Matters)*, London: Hodder & Stoughton.

Boluk, K. (2011a) 'Fair Trade Tourism South Africa: A Pragmatic Poverty Reduction Mechanism?' in *Tourism Planning & Development*, 8 (3): 237–51.

Boluk, K. (2011b) 'In Consideration of a *New* Approach to Tourism: A Critical Review of Fair Trade Tourism' in *The Journal of Tourism and Peace Research*, 1 (3): 27–37.

Briassoulis, H. (2000) 'Environmental Impacts of Tourism: A Framework for Analysis and Evaluation' in H. Briassoulis and J. van der Straaten (eds), *Tourism and the Environment Regional, Economic, Cultural and Policy Issues*, 2nd edition, Dordrecht: Kluwer Academic Publishers.

Britton, S.G. (1982) 'The Political Economy of Tourism in the Third World' in *Annals of Tourism Research*, 9 (3): 331–58.
Brown, F. and Hall, D. (2008) 'Tourism and Development in the Global South: The Issues' in *Third World Quarterly*, 29 (5): 839–49.
Budeanu, A. (2005) 'Impacts and Responsibilities for Sustainable Tourism: A Tour Operators' Perspective' in *Journal of Cleaner Production*, 13 (2): 89–97.
Butcher, J. (2003) *The Moralisation of Tourism: Sun, Sand ... and Saving the World?* London: Routledge.
Butler, R.W. (2000) 'Tourism and the Environment: A Geographical Perspective' in *Tourism Geographies: An International Journal of Tourism Space, Place and Environment*, 2 (3): 337–58.
Carlisle, S. (2010) 'Access and Marginalisation in a Beach Enclave Resort' in S. Cole and N. Morgan (eds), *Tourism and Inequality: Problems and Prospects*, Wallingford, Oxfordshire: CAB International: 67–84.
Carson, R. (1962) *Silent Spring*, New York: Fawcett Crest.
Choi, H.C. and Murray, I. (2010) 'Resident Attitudes Towards Sustainable Community Tourism' in *Journal of Sustainable Tourism*, 18 (4): 575–94.
Chong, T. (2005) 'From Global to Local: Singapore's Cultural Policy and its Consequences' in *Critical Asian Studies*, 37 (4): 553–68.
Cole, S. (2013) 'Tourism and Water: From Stakeholders to Rights Holders, and What Tourism Businesses need to do' in *Journal of Sustainable Tourism*, DOI: 10.1080/09669582.2013.776062.
Cole, S. and Morgan, N. (2010) *Tourism and Inequality: Problems and Prospects*, Wallingford, Oxford: CAB International.
Clarke, J. (1997) 'A Framework Of Approaches To Sustainable Tourism' in *Journal of Sustainable Tourism*, 5 (3): 224–33.
Diamantis, D. and Ladkin, A. (1999) 'The Links Between Sustainable Tourism and Ecotourism: A Definitional And Operational Perspective' in *Journal of Tourism Studies*, 10(2): 35–46.
D'Sa, E. (1999) 'Wanted: Tourists With a Social Conscience' in *International Journal of Contemporary Hospitality Management*, 11 (2/3): 64–8.
de Kadt, E. (1979) *Tourism: Passport to Development? Perspectives on the Social and Cultural Effects of Tourism in Developing Countries*, Oxford: Oxford University Press.
Eagles, P.F.J., McCool, S.F. and Haynes, C.D.A. (2002) 'Sustainable Tourism in Protected Areas: Guidelines for Planning and Management', Switzerland and Cambridge: IUCN Gland. Online, available at: www.cmsdata.iucn.org/downloads/pag_008.pdf (accessed 16 June 2013).
Fennell, D.A. (1999) *Ecotourism: An Introduction*, London: Routledge.
Fennell, D.A. (2006) *Tourism Ethics*, Clevedon: Channel View Publications.
Fennell, D.A. (2008) 'Tourism Ethics Needs More Than a Surface Approach' in *Tourism Recreation Research*, 33 (2): 223–4.
Goodwin, H. and Francis, J. (2003) 'Ethical and Responsible Tourism: Consumer Trends in the UK' in *Journal of Vacation* Marketing, 9 (3): 271–84.
Hall, C.M. (2010) 'Equal Access for All? Regulative Mechanisms, Inequality and Tourism Mobility' in S. Cole and N. Morgan (eds), *Tourism and Inequality: Problems and Prospects*, Wallingford, Oxfordshire: CAB International: 34–48.
Harrison, R., Newholm, T. and Shaw, D. (eds), (2005) *The Ethical Consumer*, London: Sage Publications.

Hartwick, E. (2000) 'Towards a Geographical Politics of Consumption' in *Environment and Planning*, 32: 1177–92.
Higgins-Desbiolles, F. (2010) 'Justifying Tourism: Justice Through Tourism' in S. Cole and N. Morgan (eds), *Tourism and Inequality: Problems and Prospects*, Wallingford, Oxfordshire: CAB International: 194–211.
Holden, A. and Fennell, D.A. (2013) *The Routledge Handbook of Tourism and the Environment*, London: Routledge.
Honey, M. and Gilpin, R. (2009) 'Tourism in the Developing World: Promoting Peace and Reducing Poverty', Special Report 233, United States Institute for Peace. Online, available at: www.usip.org/sites/default/files/tourism_developing_world_sr233_0.pdf (accessed 16 June 2013).
Hultsman, J. (1995) 'Just Tourism, an Ethical Framework' in *Tourism Management*, 11 (9): 553–67.
Kalisch, A. (2001) *Tourism as Fair Trade, NGO Perspectives*, London: Tourism Concern.
Kalisch, A. (2010) 'Fair Trade in Tourism: A Marketing Tool for Transformation?' in S. Cole and N. Morgan (eds), *Tourism and Inequality: Problems and Prospects*, Wallingford, Oxfordshire: CAB International: 85–106.
Kalisch, A. (2013) 'Fair Trade in Tourism: Critical Shifts and Perspectives' in A. Holden and D.A. Fennell (eds), *The Routledge Handbook of Tourism and the Environment*, Abingdon: Routledge: 494–504.
Kavallinis, I. and Pizam, A. (1994) 'The Environmental Impacts of Tourism: Whose Responsibility is it Anyway? The Case Study of Mykonos' in *Journal of Travel Research*, 33 (2): 26–32.
Kibicho, W. (2009) *Sex Tourism in Africa: Kenya's Booming Industry*, Farnham: Ashgate.
Kirtsoglou, E. and Theodossopoulos, D. (2004) '"They are Taking our Culture Away." Tourism and Culture Commodification in the Garifuna Community of Roatan' in *Critique of Anthropology*, 24 (2): 135–57.
Krippendorf, J. (1987) *The Holiday Makers: Understanding the Impact of Leisure and Travel*, Oxford: Butterworth-Heinemann.
Lisle, D. 2008, 'Joyless Cosmopolitans: The Moral Economy of Ethical Tourism', Paper presented at ISA's 49th Annual Convention, Bridging Multiple Divides, San Francisco, California. Online, available at: www.allacademic.com/meta/p254616_index.html (accessed 19 February 2010).
Mbaiwa, J.E. (2003) 'The Socio-Economic and Environmental Impacts of Tourism Development on the Okovango Delta, North-Western Botswana' in *Journal of Arid Environments*, 54 (2): 447–67.
McCabe, S. (2005) 'Who is a Tourist? A Critical Review' in *Tourist Studies*, 5 (1): 85–106.
Moufakkir, O. and Kelly, I. (2010) *Tourism, Progress and Peace*, Wallingford, Oxfordshire: CAB International.
Mowforth, M. and Munt, I. (2003) *Tourism and Sustainability: Development and New Tourism in the Third World*, 2nd edition, London: Routledge.
Mowforth, M. and Munt, I. (2009) *Tourism and Sustainability: Development, Globalisation and New Tourism in the Third World*, 3rd edition, London: Routledge.
Muganda, M., Sahli, M. and Smith, K.A. (2010) 'Tourism's Contribution to Poverty Alleviation: A Community Perspective from Tanzania' in *Development Southern Africa*, 27 (5): 629–46.

Nicholls, A. and Opal, C. (2005) *Fair Trade Market-Driven Ethical Consumption*, California: Sage Publications.

Nunkoo, R. and Ramkissoon, H. (2010) 'Gendered Theory of Planned Behaviour and Residents' Support for Tourism' in *Current Issues in Tourism*, 13(6): 525–40.

O'Hare, G. and Barrett, H. (1999) 'Regional Inequalities in the Peruvian Tourist Industry' in *Geographical Journal*, 165 (1): 47–59.

Oppermann, M. (1998) *Sex Tourism and Prostitution: Aspects of Leisure, Recreation, and Work*, New York: Cognizant Communication Corporation.

O'Reilly, C.C. (2006) 'From Drifter to Gap Year Tourist Mainstreaming Backpacker travel' in *Annals of Tourism Research*, 33 (4): 998–1017.

Peattie, K. (1992) *Green Marketing*, London: Longman.

Popke, J. (2006) 'Geography and Ethics: Everyday Mediations Through Care and Consumption' in *Progress in Human Geography*, 30 (4): 504–12.

Pritchard, A., Morgan, N. and Ateljevic, I. (2011) 'Hopeful Tourism: A New Transformative Perspective' in *Annals of Tourism Research*, 38 (3): 941–63.

Reisinger, Y. (2013) *Transformational Tourism: Tourist Perspectives*, Wallingford, Oxfordshire: CAB International.

Rutten, M. (2002) 'Parks Beyond Parks: Genuine Community-Based Wildlife Ecotourism or Just Another Loss of Land for Maasai Pastoralists in Kenya?' Issue Paper 111. Online, available at: www.opc-ascl.oclc.org (accessed 4 January 2013).

Saarinen, J. and Niskala, M. (2009) 'Selling Places and Constructing Local Cultures in Tourism: The Role of the Ovahimba in Namibian Tourism Promotion' in P. Hottola (ed.), *Tourism Strategies and Local Responses in Southern Africa*, Wallingford, Oxfordshire: CAB International: 62–72.

Sánchez-Taylor, J. (2010) 'Sex Tourism and Inequalities' in S. Cole and N. Morgan (eds), *Tourism and Inequality: Problems and Prospects*, Wallingford, Oxfordshire: CAB International: 49–66.

Sasidharan, V., Sirakaya, E. and Kerstetter, D. (2002) 'Developing Countries and Tourism Ecolabels' in *Tourism Management*, 23: 161–74.

Scheyvens, R. (1999) *Tourism for Development: Achieving Justice through Tourism*, London: Pearson.

Scheyvens, R. (2002) *Tourism for Development: Empowering Communities*, Harlow: Prentice Hall.

Scheyvens, R. (2007) 'Poor Cousins no More: Valuing the Development Potential of Domestic and Diaspora Tourism' in *Progress in Development Studies*, 7 (4): 307–25.

Scheyvens, R. and Momsen, J.H. (2008) 'Tourism and Poverty Reduction: Issues for Small Island States' in *Tourism Geographies*, 10 (1): 22–41.

Scheyvens, R. and Russell, M. (2012) 'Tourism and Poverty Alleviation in Fiji: Comparing the Impacts of Small- and Large-Scale Tourism Enterprises' in *Journal of Sustainable Tourism*, 20 (3): 417–36.

Seif, J. (2001) 'Facilitating Market Access for South Africa's Disadvantaged Communities and Population Groups through "Fair Trade in Tourism"', Fair Trade in Tourism South Africa, IUCN South Africa and University of Chicago (unpublished).

Shaw, D. and Riach, K. (2011) 'Embracing Ethical Fields: Constructing Consumption in the Margins' in *European Journal of Marketing*, 45 (7/8): 1051–67.

Shaw, D., Shiu, E. and Clarke, I. (2000) 'The Contribution of Ethical Obligation and Self-Identity to the Theory of Planned Behaviour: An Exploration of Ethical Consumers' in *Journal of Marketing Management*, 16: 879–94.

Smith, R. and Duffy, M. (2003) *The Ethics of Tourism Development*, London: Routledge.

Snyder, K.A. and Sulle, E.B. (2011) 'Tourism in Maasai Communities: A Chance to Improve Livelihoods?' in *Journal of Sustainable Tourism*, 19 (8): 935–51.

Spenceley, A. and Goodwin, H. (2007) 'Nature-Based Tourism and Poverty Alleviation: Impacts of Private Sector and Parastatal Enterprises in and Around Kruger National Park, South Africa' in *Current Issues in Tourism*, 10 (2–3): 255–77.

Szmigin, I., Carrigan, M. and McEachern, M.G. (2009) 'The Conscious Consumer: Taking a Flexible Approach to Ethical Behaviour' in *International Journal of Consumer Studies*, 33 (2): 224–31.

Tosun, C. (2002) 'Host Perception of Impact: A Comparative Tourism Study' in *Annals of Tourism Research*, 29 (1): 231–53.

Tribe, J. (2002) 'Education for Ethical Tourism Action' in *Journal of Sustainable Tourism*, 10 (2): 309–21.

Tsalikis, J. and Fritzsche, D.J. (1989) 'Business Ethics: A Literature Review with a Focus on Marketing Ethics' in *Journal of Business Ethics*, 18: 695–743.

Tucker, H. and Akama, J. (2009) 'Tourism as Postcolonialism' in J. Tazim and M. Robinson (eds), *The Sage Handbook of Tourism Studies*, London: Sage Publications: 504–20.

UNEP (2011) 'Towards a Green Economy: Pathways to Sustainable Development and Poverty Eradication: A Synthesis for Policy Makers', United Nations Environment Programme, Nairobi, Kenya. Online, available at: www.unep.org/greeneconomy (accessed 20 April 2013).

Wearing, S. (2001) *Volunteer Tourism: Experiences that Make a Difference*, Wallingford, Oxfordshire: CAB International.

Wearing, S. (2004) 'Examining Best Practice in Volunteer Tourism' in R. Stebbins and M. Graham (eds), *Volunteering as Leisure/ Leisure as Volunteering: An International Assessment*, Wallingford, Oxfordshire: CAB International: 209–24.

Weeden, C. (2002) 'Ethical Tourism: An Opportunity for Competitive Advantage?' in *Journal of Vacation Marketing*, 8 (2): 141–53.

Weeden, C. (2013) *Responsible Tourist Behaviour*, London: Routledge.

Wheeller, B. (1991) 'Tourism's Troubled Times: Responsible Tourism is Not the Answer' in *Tourism Management*, 12 (2): 91–6.

Wilkins, L. (1999) 'The New Millennium: Linking Environmental and Social Issues in Fair Trade' in *Networks*, Spring Issue.

Wright, J. (2006) 'Code Green' in *The Australian*, 29 April.

Part I
Debates on ethical consumption in tourism

2 What does it mean to be good in tourism?

Kellee Caton

Introduction

In an essay written around the time he entered the last decade of his life, evocatively titled "Trotsky and the Wild Orchids," the world-renowned philosopher Richard Rorty (1999) paused to reflect in personal terms on what had become one of the most central philosophical questions driving his 40-year career: how to reconcile the tension between pursuing one's own idiosyncratic personal passions in life, on one hand, and living a life of good citizenship on the other. Rorty writes of his days as a young boy, running errands for his parents, who were active in the New York Trotskyite movement in the late-30s and early-40s. While in transit on the subway, he would read the documents he was delivering, from which he learned much about the struggles being faced by society's underdogs, including union organizers, sharecroppers and black American workers, as they struggled for better working conditions and opportunities in an era of industrial and corporate strong-arming.

By the age of 12, he recalls, his observations of the work done by those in his parents' social circle had cemented in him an awareness that "the point of being human was to spend one's life fighting social injustice" (Rorty 1999: 6). At the same time, however, the young Rorty also had other interests, most notably a passion he recalls for finding wild orchids in the woods of the American Northeast, and then pouring over musty botany books at the public library in an attempt to categorize his findings from among the 40 species that grew in the area. He had a nagging sense of guilt, however, that there was something "a bit dubious" about this esoteric hobby, and he felt that Trotsky would somehow not have approved of him wasting his time trying to track down and categorize these "socially useless flowers" (Rorty 1999: 7). When he became old enough, Rorty set off to study philosophy in the hope of understanding how these two sides of human existence – the drive to pursue personal, idiosyncratic, self-fulfilling passions and the desire for meaningful public engagement to improve the lives of others – could be reconciled. After decades of consideration, he finally decided that the trick couldn't be done – that there was no way to synthesize the work of philosophers like Nietzsche, Foucault, Heidegger and Kierkegaard, who speak a language of autonomy and self-creation, with that of thinkers like Marx, Dewey,

Mills and Habermas, who are primarily concerned with human solidarity and collective social improvement (Rorty 1989, 1999).

Many of us who love to travel face a similar quandary. On one hand, tourism is about freedom, relaxation and escape – a chance to indulge in fantasy and experience exotic and invigorating sensations that provide a contrast from the stressors, routines, and *ennui* that often characterize regular workaday life. We travel to experience pleasure and to satisfy our own personal desires (Butcher 2003). On the other hand, however, tourism is also a morally loaded territory. It is a practice through which we imagine and encounter people, places and cultures beyond our own everyday life space (Hollinshead 2007; Caton 2012). This practice is consequential, in both discursive and material terms, and its moral weight is finally beginning to receive public awareness – part of a larger awakening in popular culture to the social and environmental fallout that is ensuing from several decades of unbridled consumerism in the Western world, in which the sheer level of production and consumption of material goods in society far outstrips that of any earlier point in human history (Schor and Holt 2000; Glickman 1999; Cross 2000; Cohen 2003). The growing awareness of consumerism's consequences is occurring both in general (Schor 2000; Moore 2007; Littler 2009; Humphrey 2010), and more specifically within the context of tourism (Curtin and Busby 1999; Cleverdon and Kalisch 2000; Weeden 2002), as today's tourists are increasingly beginning to face up to the fact that their behaviors have ramifications for others. This issue thus becomes one more to be grappled with in the ongoing construction of an ethical self.[1]

The realization that our consumption practices can have negative ramifications for other people, and for the health of our planet, has sparked an increasing desire to do something about this problem, and the primary response has come in the form of an interest in "ethical consumerism"[2] – the practice of "voting with one's wallet," or making purchasing decisions that one feels will have a better impact on society or nature. Although the term "ethical consumerism" may be relatively new in academic circles, becoming popularized around the beginning of the 1980s, the idea has been around a long time. Irving *et al.* (2002) cite the beginning of the South African boycott in 1959 and the publication of a "green consumer guide" in 1971 as early instances of what we tend to think of as the modern ethical consumerism movement, but they note examples of anti-imperialist boycotts in China in 1905, US labor movement boycotts at the beginning of the twentieth century, and even anecdotal accounts of sugar boycotts against plantations riding on the back of slave labor in the nineteenth century, to illustrate the long history of ethical consumerism thought and practice.

Today's ethical consumerism primarily takes the form of supporting particular merchants and eschewing the wares of others, based on evidence or perceptions about such merchants' social and environmental practices. For example, a consumer may consider such issues as how particular products are manufactured, what sorts of resources are used, who is employed in production, what labor conditions characterize production, and how waste is dealt with. Similarly, concerns about ethical consumption in the context of tourism have largely taken the form of an interest in the social and environmental practices of tourism service providers, such as tour operators, hotels, restaurants and transportation providers (Weeden 2002). Current discourses and practices of ethical consumption in tourism, however, constitute only one way of imagining what it means to "be good" in tourism, and viewing moral action in tourism through the lens of consumption, although not without great benefits to the world, runs the risk of obscuring other, non-consumerist frameworks for imagining what the tourism encounter is about and how we might shape that encounter into a space that better reflects our moral goals. In this brief conceptual chapter, I seek to provide some historical context for how we have come to construct ourselves as ethical subjects within the contemporary consumer-oriented cultural milieu, and then to reflect on the consequences of this consumerist ethical subjectification for achieving moral action in tourism in terms of the potentialities for critical action that it opens up and those that it leaves in the shadows.

The new consumer self

Although people have always consumed material items, and such items have long figured heavily in the construction of individual and community identity (Trentmann 2009), the sheer pace and volume of consumer activity in the contemporary world, and its concomitant impact on identity-building, has generally led sociocultural theorists to conclude that we are living in a distinct historical moment characterized by a *culture* of consumerism. In many ways, one can look to the Industrial Revolution and its ensuing progeny, Fordism, for the birth of this consumer culture. The mass production capabilities arising through the Industrial Revolution led to a situation in which, for the first time in history, the supply of goods being manufactured could vastly outstrip any demand for those goods which had a basis in people's real needs (Gabriel and Lang 2008). The captains of industry were thus faced with an historically new problem: they needed not only to manufacture goods, but also to manufacture markets to consume those goods. This task was answered partially by the Fordist Deal, which "recognized the potential of ... workers as customers" if they were paid above the subsistence level and issued the "promise of ever-increasing standards of living in exchange for a quiescent labour force accepting alienating work" (Gabriel and Lang 2008: 323). It was also answered partly by the advertising industry, which was charged with convincing the public to desire – or even to perceive a *need* for – the goods that were being produced, instead of the other way around (Cross 2000; Barber 2007).

The marketing process was perhaps fairly innocuous at first, with most advertising focusing on conveying more-or-less objective facts about the product at hand (although there certainly were instances of tall tales in making claims about the efficacy of products, especially in the pharmaceuticals industry, which was not regulated in the United States or Canada until the early twentieth century) (Jackson Lears 1983). But in the post-war economic boom of the 1950s and beyond, the amount of disposable income to hand in the Western world signaled too great an opportunity for the commercial sector, which responded by producing goods at a never-before-seen pace and volume (Cross 2000). This output was accompanied by increasingly aggressive marketing activities, which continue unabated to the present day. Consumption was also harnessed to political life in new ways, as consuming came to be articulated as a veritable act of patriotic citizenship, with citizens being urged to "do their part" by spending, rather than saving, in order to keep "the economy" pumping (Cohen 2003).

In recent decades the process has gone global, an outcome partially due to the Fordist ideal having been exported through the outsourcing of industry. Products could be made more cheaply in so-called developing countries, and rising incomes among workers in these countries could then set the condition for the rise of yet another pocket of consumers. Given its extraterritorial nature, global capitalism today shows little sign of being restrained in any meaningful way, as there is no supranational political entity with the power to police it through regulation or taxation (Barber 2007). It creates, as it surges forth, a global class of consumer "haves" and a class of "have nots" – failed consumers, whose existence does not "contribute" to the global economy and who hence become the "human waste" of the system (Bauman 1998).

All of this is not to say that individual human agency has played no role in the creation and maintenance of a global consumer society. Certainly, individuals think for themselves and make choices; they are not mere dupes of a sophisticated marketing system. It is also not to say that sociocultural forces are the only powers being brought to bear on this situation. People are driven by complex biological, as well as social, forces that exert pressure on cognition and behavior, and indeed biological and social tendencies can interact to reinforce one another. Research in evolutionary psychology illustrates that, as part of our existence as social creatures, we are wired to be attuned to distinctions in status, and indeed to pursue status and the favorable opinions of others in our social group (Wright 1994; Pinker 2002) – a predilection that culture sometimes encourages.[3] Critiques of consumer culture implicitly take this into account all the way back to Veblen (1899), with his amusing skewering of the well-to-do and their propensity for "conspicuous consumption" as a way of showing off in front of friends and neighbors. We are also, as Boyd (2009) has argued, wired to be artistic creatures, our tendency toward art-making having favored our species in numerous ways throughout evolutionary history – and culture has certainly supported this biological tendency, at least in most cases.

To the degree that consumer products beyond life's necessities can aid individuals in the pursuit of status or serve as raw materials in our quest for artistic self-expression (not to mention bringing us direct physical or mental pleasure), then it is rational that we would view consuming at high levels to be in our self-interest, assuming we were not taking a longer-term view of the consequences that included knowledge of the environmental impacts of consumption (Taylor and Tilford 2000) or of what is actually a negative association between materialism and human happiness (Kasser 2002). Consumer compliance notwithstanding, however, it is unlikely that organizations would spend such a massive amount of capital on advertising expenses if the process were ineffective, so it is probably still reasonable to assign a good bit of the cause of the rise of modern consumer society to those on the supply side of the equation, who are happy to harness convenient biological and social drives to make consuming at high levels even more attractive than it would have been in the first place.

As scholars of consumer culture generally argue, the combination of the onslaught of production and marketing over the last several decades in the Western world (and increasingly in the world at large), and consumers' positive response to this onslaught, has done more than simply separate the average worker from his or her spending money. There have been greater consequences, in terms of human beings' own identities and relationships to themselves, wrought from this turn of events. To understand this situation it is helpful to borrow some conceptual footing from late-twentieth-century continental philosophy concerning the process of human identity-building.

Perhaps largely because of the long influence of Christianity over Western thought, much of the history of Western philosophy has been grounded in a belief in the existence of a human soul and in a conceptualization of individual human identity as an expression of that soul and its unique, innate characteristics. Increasingly, since the Enlightenment, and especially in the latter part of the twentieth century, however, philosophers have worked to posit, as Nietzsche put it (Cox 1999), a new version of the "soul-hypothesis," in which identity comes to be seen more as a process than as property under the ownership of an individual human. Examples of this sea change in the body of scholarly theorization are numerous. Nietzsche broke ground in this direction by conceptualizing the self as multiplicit rather than unified – as a bundle of disparate perspectives, instincts, desires, and so forth, which is reactive to the world around it (Cox 1999). Ricoeur (1992) refined this idea and posited a notion of the self as a contingent, temporal entity which comes into being at the intersections of the "fault lines" between what has been done to it, beyond its control, and what it does, through its own agency. With an emphasis on the power of language and discourse to create social reality, Foucault directed focus to the construction of individual identity as a profoundly social process, and it is his work that is perhaps most valuable for understanding the rise of the consumer self.

In what was arguably his most original contribution to social theory (Rabinow 1984), Foucault coined the notion of "subjectivation" (frequently referred to by scholars working in his wake as "subjectification," which perhaps flows better as an English translation). In our contemporary culture of "self help" and "personal growth," it is easy to imagine that our identities are a feature of ourselves that we own, and that their shaping is something that unfolds more-or-less under our own direction. But Foucault took the opposite stance, arguing in line with the philosophers mentioned above that identity is produced as a complex response to various external forces, such as social norms or access to particular resources and opportunities, and that these forces are historically contingent. We are typically not wise to this process, however. Instead, we tend to internalize the identities expressed through the appearances, behaviors, and other communicative acts that we produce at the behest of the social norms that govern our interactions with others, and this constitutes our subjectivity – our historically contingent and relationally determined selfhood. Key to the idea of subjectification, then, is the process through which we buy in, on an individual psychological level, to our socially produced identity, such that it becomes a compelling description *of* ourselves *to* ourselves (Foucault, under pseudonym Florence, 1984).

Foucault explored the notion of subjectification predominantly within the contexts of sexuality (1988a, 1990a, 1990b) and sanity (1988b, 1994),[4] but it clearly has many other applications, including one relevant for our current purpose: the self as consumer. Before the mass production of goods, and the widespread ability of inhabitants of wealthy parts of the world to consume them, people tended to define their lives with reference to a variety of contexts in which they were enmeshed, including work, family, religion and community, along with the material goods that featured in their lives. In the United States the "Puritan work ethic," in which the very act of leading a morally upright life was directly bound up with one's propensity to pursue productive activities in one's time on earth, for instance, stands as a testament to the earlier influence of forces like work and religion in people's sense of self-definition. Although it is still the case today that identity is the product of interaction with complex life contexts, there has been a significant shift in the relative importance of the products we consume in defining our identities, such that, in many ways, "we are what we buy" (or at least we think we are) (Campbell 2004).

This shift has not been a random result of the dramatic increase in the availability of affordable goods, but rather represents a concerted effort on the part of marketers to infuse the purchasing process with an emotional dimension, leveraging what some might call people's spiritual or humanist capacities (such as the ability to feel affection or compassion) in order to generate consumer interest in and loyalty to particular products and brands.[5] Barber (2007) provides a superb example of this propensity in his discussion of Saatchi & Saatchi CEO Kevin Roberts's notion of "lovemarks" (also the title of Roberts's book), a twenty-first century replacement for the old-school notion of "brands" or "trademarks," which according to the likes of Roberts have gone frumpy and are in dire need of a face-lift. As Barber (2007: 183–4) explains,

> In [*Lovemarks*, Roberts] is happy to recognize marketing not as capitalism's tool but as its new essence – emotional propaganda that hijacks authentic emotions and sentiments and employs them in wholly instrumental ways to sell products to which neither producers nor consumers (absent Kevin Roberts's stratagems) otherwise are likely to have much interest, and for which there is in any case little inherent demand. Roberts believes that the time has come for a third-generation term for *trademarks*, one which improves on *brands* in the same way *brands* once improved on *trademarks*. His candidate is *lovemarks* – brands that consumers are in love with, and hence which are "owned" not by corporate managers or stockholders but by consumers themselves. To put it more plainly, lovemarks are brands with which

consumers can be made to fall in love and then persuaded to "own" by savvy advertisers ... [whose] job is to immerse products and services in a nonspecific sentimental miasma from which "emotional decisions" can "naturally" arise (natural as a creation of artifice, and emotional decisions as irrational and nondeliberative and hence scarcely decisions at all). It's "love for sale" in the most literal sense.

Industry also has its fingers in identity in a direct way, through the increasing exploitation of categories of selfhood like ethnicity, sexuality, and "lifestyle," as ever-advancing market saturation generates the increasing need to manufacture "niche products," to find new markets for the ever-more highly differentiated raw materials of identity in which producers and marketers deal (Giroux 1993; du Gay 1996). Race and ethnicity, in particular, has become an increasingly frequently tapped well for fueling this production of niche goods, as the popularity of everything from world music to "third world cinema" to ethnically inspired fashion readily attests. In the words of Davidson (1992: 199), "capital has fallen in love with difference." Consumer products provide a perceived avenue for self-definition within the bounds of a particular sociodemographic or lifestyle category, as well as an opportunity to explore and experiment with other ways of imagining oneself.

The pervasiveness of consumerism as a template for identity construction has been well documented in conceptual terms, with scholars like Campbell (2004: 41–2) remarking evocatively that as "more and more areas of contemporary society have become assimilated to a 'consumer model' ... the underlying metaphysics of consumerism has in the process become a kind of default philosophy for all modern life." It has also been expressed in concrete terms, as the target of frustration of individuals in particular professions, such as higher education or medicine, who are seeing the socioscape of their occupational interactions change; what was once a relationship between two parties (e.g. teacher and student, doctor and patient) based on trust, expertise, and care, has now, having been transmuted through the distorting lens of consumer culture, become an exchange between a salesperson and a customer (Gabriel and Lang 2008). It should not be any surprise then that when we seek to construct an ethical self our consumer-inflected identities end up providing the raw material.

The ethical subject on tour

Ethical consumerism has not been without its victories. Irving *et al.* (2002) detail many concrete successes of ethical consumer action, which include effecting positive change in the areas of animal welfare, the environment, fair trade and workers' rights, and human rights more broadly construed. Tourism, however, is, if not a completely unique, then at least a highly unusual arena of economic exchange. The commodities of tourism are not primarily widgets produced by undercompensated workers in some destination country factory – although the souvenir kitsch consumed in vast quantities by tourists certainly bespeaks such an aspect of material production and consumption in tourism, and comes with its own set of moral problems from a Marxist commodity fetishism standpoint. Instead, what is mostly bought and sold in tourism is some combination of fantasy and experience, a jointly discursive and embodied "product" that draws for its raw materials on the geographic spaces, the cultural dynamics, and even the physical bodies of other human beings.

The problem, then, with taking an ethical consumerist approach to morality in tourism is not that there is nothing to be gained by tourists becoming more informed about the social and environmental impacts of various purchasing decisions while traveling, but rather that the consumer subjectivity so pervasive today is simply not equipped to deal with the challenges of human interaction in tourism. A consumer subjectivity inherently flattens human beings, reducing our relationships to the role of parties in an economic exchange. Many of the moral conundrums of tourism, however, exist on an entirely different plane, involving questions, for example, about rights to space and particular uses of it, rights of representation and narration of meaning, and rights and responsibilities associated with the interpersonal interactions of performing and gazing.

Indeed, the reduction of these complex dynamics to a commodity framework is directly responsible for so many of the injustices, big and small, that we witness every day in contemporary tourism practice. These range from the mildly annoying and laughable, as in the case of the ethnocentric tourist who turns down an invitation to attend an event that holds deep spiritual meaning to a local community because he "has already seen the show" in the dinner theatre of his hotel lobby (Bruner 2005); to the more deeply problematic, as exemplified by residents of destination countries who do not have access to their own beaches because those spaces have been cordoned off as resort enclaves for generating multinational corporate profit (Carlisle 2010); to the truly tragic, when the bodies of non-consenting teenagers are offered up for cash for the sexual gratification of tourists acting out their colonialist fantasies (Ryan and Hall 2001; Taylor 2010). Even independent travel, frequently represented through the ideal of the backpacker tourist, is increasingly taking on commercial overtones, as spaces in destination countries become reconfigured to cater to backpackers' needs in ways that render them unwelcoming to locals and domestic visitors (Teo and Leong 2006; Wilson et al. 2008).

And yet, because our identities are so imbricated with consuming, and because we perceive a lack of other viable possibilities for political engagement in the face of global capitalism's transcendence of any sort of really binding framework of legal authority (because, as noted, no supranational form of government exists on a scale sufficient to police the global market on behalf of the public interest) (Barber 2007), we end up redrawing the lines of citizenship to emphasize political action through the market. It is courtesy of this process that we see strange new forms in tourism, such as the phenomenon of touring "slums" in India, Brazil or South Africa, in order both to gaze and to feel like one is making a financial contribution, through one's paying to participate, to the economic improvement of the lives of those who dwell there – a practice Leite and Graburn (2009: 55) have characterized as raising questions about the "commodification ... of moral outrage." Thus, even as we try to heed the call of scholars like Higgins-Desbiolles and Blanchard (2010) to think outside the binds of neoliberal marketization and remember that tourism should also be contextualized in relationship to such things as human rights, conflict resolution, justice and peace, we continue to find ourselves falling back into market frameworks to deal with complex moral goals, like trying to comprehend the suffering or the sense of purpose of another.

Conclusion

Ethical consumerism is one way of constructing an ethical self, and one that exhibits increasing popularity in the contemporary moment. Insofar as consumer exchange is one of the key modes today through which we relate to our fellow humans, ethical consumerism has its uses; it can help us to better navigate the space of "shoulds" in the inevitable purchasing decisions we make in travel and in daily life. In practice, it has seen its share of victories, such as those mentioned above, and also its share of failures, including "greenwashing," "conscience fatigue," and consumer failure to echo stated preferences for ethically produced goods and services upon actually reaching the cash register (Irving *et al.* 2002; Devinney *et al.* 2010). In principle, as a discourse, it has arguably been even more valuable, serving to induce a kind of reflexivity in consumers (even if they do not always yet act) that is helping to shatter the heady sense of freedom from consequence regarding overconsumption and waste that has characterized Western popular culture for decades.

But commerce is not the only ground where we meet one another as human beings. Many of the problems of contemporary tourism, given so much attention by scholars and industry partners of late, exemplify the challenges that arise when we forget this fact – when more and more domains of social life and interaction begin to tumble under the spectre of commoditization. There are many other problems of a moral order in tourism that are typically neglected in discussions in the literature, presumably at least partially because we cannot "buy answers" for them. They are more complex than problems of economic exchange. They hint at deeper challenges inherent in the human predicament of being both individual and social. They need philosophical attention, and this is the work of fully engaged minds, alive with moral imagination – not the work of invisible hands.

It's *hard* to grapple with the idea of moral action in tourism because so many of the major motivations to travel are firmly centered in satisfying the needs and desires of the self. We frequently travel for pleasure, escape, rest and recuperation, self-experimentation and personal development. When we do so, however, we use other people and their resources. We help to reshape the physical landscapes of others' spaces. We stumble around in their cultural politics. We consume their bodies, their cultural inheritances, the outputs of their creativity. We use them as objects in creating our stories of self.

Just as Rorty concluded, we will always live with the tension between nurturing the self and doing right by others. Constructing an ethical self in tourism ultimately has – or at least should have – more to do with how we decide to navigate that tension as a whole than with how we behave in any given moment of commercial exchange. To view the construction of an ethical self in tourism solely, or even primarily, through the lens of ethical consumption is to engage in one more form of economic reductionism – to draw us deeper into a consumer subjectivity that reconfigures ever-more aspects of life in material transactionist terms. Human life is bigger than that, though, and so we need to be asking bigger questions about how we can be good in tourism.

Activity

Thought exercise: the other side of the tourism encounter

Tourism students often find it easy to imagine what it is like to be a tourist; after all, it's a role most of us have played at some point in our lives, and often a very enjoyable one! It is sometimes harder, however, to put ourselves in the shoes of those who live and work in tourist destinations. Such people find their spaces and cultural practices turned into commodities, a process that can lead to mixed emotions.

In order to better understand the position that people living in tourist destinations find themselves in, as well as to explore some of the ethical complexities of tourism discussed in this chapter, please reflect on the following questions and then discuss them in small groups with your classmates:

1 Imagine that a group of tourists from another country wanted to come to your hometown to experience your culture. What aspects of your town and your life would you want to share with them?
2 How would this sharing process make you feel?
3 How would you feel if this group of tourists wanted to pay you for what you were sharing with them?
4 Can you think of anything about your life or culture that you wouldn't feel comfortable sharing with tourists?
5 If so, imagine that this is what tourists most wished they could see. Are there any circumstances under which you could be persuaded to share it? Would it make it better or worse if the tourists offered to pay you to share it?

Suggested reading

Barber, B. (2007) *Consumed: How Markets Corrupt Children, Infantilize Adults, and Swallow Citizens Whole*, New York: W.W. Norton and Company.
Du Gay, P. (1996) *Consumption and Identity at Work*, London: Sage.
Schor, J. and Holt, D. (eds) (2000) *The Consumer Society Reader*, New York: The New Press.

Notes

1 None of this, however, is to suggest that this movement represents the views of anything like the majority of tourists. Although, as Weeden (2002) argues, it is increasingly being recognized that most purchasing decisions have an ethical dimension, studies by tourism scholars have found that tourists may have little understanding of the social and environmental impacts of tourism, and little imagination regarding how their own behaviors as tourists could change things for the better (Miller *et al.* 2010).
2 Ethical consumerism is not an uncontested concept, with some even arguing that the "ethical consumer" is an academic myth (Devinney *et al.* 2010), although this appears to be a minority opinion.
3 It would be an error, however, to claim that culture always encourages its members in their status-seeking behaviors. Indeed, many cultural settings and products, from Kindergarten classrooms to state social welfare policies, encourage egalitarianism, which can be important for group cohesion and which is also supported at the evolutionary-biological level by the drive for collaboration (Wright 2000; Boyd 2009).
4 But see also *The Order of Things* (1994b) and *Ethics: Subjectivity and Truth* (1997) for more general discussions.
5 In fact, this has become such a trope in advertising that some companies are actively seeking to subvert it and to exploit the potential that lies in satirizing it; consider, for instance, the popular Ikea lamp commercial directed by Spike Jonze (and still viewable on YouTube), which pokes fun at consumers' propensity to invest their material possessions with feelings and personalities and to form emotional relationships with them.

References

Barber, B. (2007) *Consumed: How Markets Corrupt Children, Infantilize Adults, and Swallow Citizens Whole*, New York: W.W. Norton and Company.
Bauman, Z. (1998) *Globalization: The Human Consequences*, New York: Columbia University Press.
Boyd, B. (2009) *On the Origins of Stories: Evolution, Cognition, and Fiction*, Cambridge, MA: Belknap Press.
Bruner, E. (2005) *Culture on Tour: Ethnographies of Travel*, Chicago: University of Chicago Press.
Butcher, J. (2003) *The Moralisation of Tourism: Sun, Sand ... and Saving the World?*, London: Routledge.
Campbell, C. (2004) "I shop therefore I know that I am: The metaphysical basis of modern consumerism" in K. Eckstrom and H. Brembeck (eds.), *Elusive Consumption*, Oxford, UK: Berg: 27–44.
Carlisle, S. (2010) "Access and Marginalization in a Beach Enclave Resort" in S. Cole and N. Morgan (eds.), *Tourism and Inequality: Problems and Prospects*, Oxfordshire, UK: CAB International: 67–84.
Caton, K. (2012) "Taking the Moral Turn in Tourism Studies" in *Annals of Tourism Research*, 39: 1906–28.
Cleverdon, R. and Kalisch, A. (2000) "Fair Trade in Tourism" in *International Journal of Tourism Research*, 2: 171–87.
Cohen, L. (2003) *A Consumers' Republic: The Politics of Mass Consumption in Postwar America*, New York: Alfred A. Knopf.
Cox, C. (1999) *Nietzsche: Naturalism and Interpretation*, Berkeley: The University of California Press.

Cross, G. (2000) *An All-Consuming Century: Why Commercialism Won in Modern America*, New York: Columbia University Press.

Curtin, S. and Busby, G. (1999) "Sustainable Destination Development: The Tour Operator Perspective" in *International Journal of Tourism Research*, 1: 135–47.

Davidson, M. (1992) *The Consumerist Manifesto*, New York: Routledge.

Devinney, T., Auger, P. and Eckhardt, G. (2010) *The Myth of the Ethical Consumer*, Cambridge: Cambridge University Press.

du Gay, P. (1996) *Consumption and Identity at Work*, London: Sage.

Foucault, M., using pseudonym Florence, M. (1984) "Foucault entry in *Dictionnaire des philosophes*," available in English translation online at: http://foucault.info/foucault/biography.html.

Foucault, M. (1988a) *The History of Sexuality, Volume 3: The Care of the Self*, New York: Vintage Books.

Foucault, M. (1988b) *Madness and Civilization: A History of Insanity in the Age of Reason*, New York: Vintage Books.

Foucault, M. (1990a) *The History of Sexuality, Volume 1: An Introduction*, New York: Vintage Books.

Foucault, M. (1990b) *The History of Sexuality, Volume 2: The Use of Pleasure*, New York: Vintage Books.

Foucault, M. (1994a) *The Birth of the Clinic: An Archaeology of Medical Perception*, New York: Vintage Books.

Foucault, M. (1994b) *The Order of Things: An Archaeology of the Human Sciences*, New York: Vintage Books.

Foucault, M. (1997) *Ethics: Subjectivity and Truth*, New York: The New Press.

Gabriel, Y. and Lang, T. (2008) "New Faces and New Masks of Today's Consumer" in *Journal of Consumer Culture*, 8: 321–40.

Giroux, H.A. (1993) "Consuming Social Change: The 'United Colors of Benetton'" in *Cultural Critique*, 26: 5–32.

Glickman, L. (ed.) (1999) *Consumer Society in American History: A Reader*, Ithaca, NY: Cornell University Press.

Higgins-Desbiolles, F. and Blanchard, L. (2010) "Challenging Peace Through Tourism: Placing Tourism in the Context of Human Rights, Justice and Peace" in O. Moufakkir and I. Kelly (eds.), *Tourism, Progress and Peace*, Wallingford, Oxfordshire: CAB International: 35–47.

Hollinshead, K. (2007) "'Worldmaking' and the Transformation of Place and Culture: The Enlargement of Meethan's Analysis of Tourism and Global Change" in I. Ateljevic, A. Pritchard, and N. Morgan (eds.), *The Critical Turn in Tourism Studies: Innovative Research Methodologies*, Amsterdam: Elsevier: 165–96.

Humphrey, K. (2010) *Excess: Anti-Consumerism in the West*, Cambridge: Polity Press.

Irving, S., Harrison, R. and Rayner, M. (2002) "Ethical Consumerism: Democracy Through the Wallet" in *Journal of Research for Consumers*, 3. Online, available at: www.jrconsumers.com/academic_articles3/issue_3/?f=5789 (accessed December 9, 2012).

Jackson Lears, T. (1983) "The Rise of American Advertising" in *The Wilson Quarterly*, 7: 156–67.

Kasser, T. (2002) *The High Price of Materialism*, Cambridge, MA: The MIT Press.

Leite, N. and Graburn, N. (2009) "Anthropological Interventions in Tourism Studies" in T. Jamal and M. Robinson (eds.), *The Sage Handbook of Tourism Studies*, London: Sage: 35–64.

Littler, J. (2009) *Radical Consumption: Shopping for Change in Contemporary Culture*, New York: Open University Press.
Miller, G., Rathouse, K., Scarles, C., Holmes, K. and Tribe, J. (2010) "Public Understanding of Sustainable Tourism" in *Annals of Tourism Research*, 37: 627–45.
Moore, A. (2007) *Unmarketable: Brandalism, Copyfighting, Mocketing and the Erosion of Integrity*, New York: The New Press.
Pinker, S. (2002) *The Blank Slate: The Modern Denial of Human Nature*, New York: Viking.
Rabinow, P. (ed.) (1984) *The Foucault Reader*, New York: Pantheon Books.
Ricoeur, P. (1992) *Oneself as Another*, Chicago: The University of Chicago Press.
Rorty, R. (1989) *Contingency, Irony, and Solidarity*, Cambridge: Cambridge University Press.
Rorty, R. (1999) *Philosophy and Social Hope*, London: Penguin Books.
Ryan, C. and Hall, C. (2001) *Sex Tourism: Marginal People and Liminalities*, London: Routledge.
Schor, J. (2000) "The New Politics of Consumption" in J. Cohen and J. Rodgers (eds.), *Do Americans Shop too Much?* Boston: Beacon Press: 3–33.
Schor, J. and Holt, D. (eds.) (2000) *The Consumer Society Reader*, New York: The New Press.
Taylor, B. and Tilford, D. (2000) "Why Consumption Matters" in J. Schor and D. Holt (eds.), *The Consumer Society Reader*, New York: The New Press, 463–87.
Taylor, J. (2010) "Sex Tourism and Inequalities" in S. Cole and N. Morgan (eds.), *Tourism and Inequality: Problems and Prospects*, Wallingford: Oxfordshire: CAB International: 49–66.
Teo, P. and Leong, S. (2006) "A Postcolonial Analysis of Backpacking" in *Annals of Tourism Research*, 33: 109–31.
Trentmann, F. (2009) "Crossing Divides: Consumption and Globalization in History" in *Journal of Consumer Culture*, 9: 187–220.
Veblen, T. (1899) *The Theory of the Leisure Class*, New York: Macmillan.
Weeden, C. (2002) "Ethical Tourism: An Opportunity for Competitive Advantage?" in *Journal of Vacation Marketing*, 8: 141–53.
Wilson, J., Richards, G. and MacDonnell, I. (2008) "Intracommunity Tensions in Backpacker Enclaves: Sydney's Bondi Beach" in K. Hannam and I. Ateljevic (eds.), *Backpacker Tourism: Concepts and Profiles*, Clevedon: Channel View Publications: 199–214.
Wright, R. (1994) *The Moral Animal: Why We Are the Way We Are: The New Science of Evolutionary Psychology*, New York: Vintage Books.
Wright, R. (2000) *Nonzero: The Logic of Human Destiny*, New York: Vintage Books.

3 You can check out anytime you like but you can never leave

Can ethical consumption in tourism ever be sustainable?

C. Michael Hall

Introduction

Ethical consumption is not new. A desire for more ethically appropriate consumption patterns, or consumption that has an underlying ethical foundation has, in the West, roots in Quakerism, Transcendentalism and even elements of Puritanism (Shi 1986), while in a more modern form it also finds expression in the counter cultures of the 1960s (Musgrove 1974). More recently, it also has found expression in the notion of voluntary simplicity (Shaw and Newholm 2002), which refers to 'the choice out of free will (rather than being coerced by poverty, government austerity programmes, or being imprisoned) to limit expenditures on consumer goods and to cultivate nonmaterialistic sources of satisfaction and meaning' (Etzioni 2003: 7). Moreover, there is also a wealth of research interest in the implications of consumer action with respect to environmental, political and social causes dating back to the 1960s and 1970s (Hall 2011a). For example, Webster (1975: 188) defined the socially conscious consumer as 'a consumer who takes into account the public consequences of his or her private consumption or who attempts to use his or her purchasing power to bring about social change'. Roberts (1993: 140) expanded the notion of conscious consumption to refer to the socially responsible consumer as 'one who purchases products and services perceived to have a positive (or less negative) influence on the environment or who patronizes businesses that attempt to effect related positive social change'.

Historically, the ethicality of consumption was often tied in to religious dictates and values over what could and could not be consumed. Such constraints on consumption operate to the present day in a number of significant consumer markets, for example in the Islamic faith with its belief in notions of *halal* (permissible) and *haram* (forbidden) products and services (Syed Marzuki *et al.* 2012). In a tourism and hospitality context this was usually expressed with respect to the desire to travel as part of a spiritual pilgrimage in which the journey was often as important as the destination (Hall 2006). However, in a modern secular context, while the act of journeying can still have significant spiritual dimensions, the focus has shifted to a more conscious awareness of what is being consumed and why. In the case of travel, this may be expressed as not travelling so far or not travelling at all for environmental reasons, or travelling to specific locations with the belief that consumption will bring social, economic and/or environmental benefits. Such is the interest in conscious consumption that tourist and hospitality products and promotions are increasingly focusing on environmental and other benefits of purchase (Peattie and Collins 2009).

But what then is ethical in consumption? A number of different dimensions may be identified (Vitell 2003; Newholm and Shaw 2007).

- appropriate and inappropriate purchasing and shopping behaviours (Vitell 2003)
- consumer resistance and activism (Sen *et al.* 2001)
- consumption morality in relation to concerns over sustainability (Connolly and Prothero 2003)
- corporate and social entrepreneurial attempts to promote ethical consumption opportunities; what is sometimes referred to as corporate social responsibility and societal marketing (Mohr *et al.* 2001)
- ethical consumption as a conscious political project of individuals and small groups as part of a broader politics of consumption (Cohen 2006).

One of the earliest definitions of ethical consumption came from Webster (1975: 188), who defined the socially conscious consumer 'as a consumer who takes into account the public consequences of his or her private consumption or who attempts to use his or her purchasing power to bring about social change'. Yet such a definition raises a number of significant issues with respect to the environmental or sustainable dimensions of consumption. For example, Cooper-Martin and Holbrook (1993) found that many US consumers regarded the purchase of American products as an ethical consumption act for nationalistic rather than environmental reasons. Nevertheless, such an understanding of the morality of consumption does reinforce the significance of a growing trend to understand ethical consumption not just in terms of the act or experience of consumption and the cost and utility of what they consume, but also in terms of the ways in which they are produced, processed and transported (Trentmann 2007). Through ethical consumption, Carrier (2010: 672) argues,

> people can do two things. Firstly, at a personal level, they can lead lives that are more moral. Secondly, at a public level, they can use their purchases to affect the larger world, by putting pressure on firms in a competitive market to change the ways that they do things.

At first glance the development of ethical consumption in tourism would seem to be something to be welcomed. By any empirical measure tourism is less sustainable than ever, with its absolute contribution to climate change and other forms of environmental change increasing each year (Hall 2011b; Scott *et al.* 2012a). At one level this can be explained by the sheer growth in tourist numbers. Even with hoped-for per trip efficiency gains the absolute contribution of tourism to climate change will continue to increase in the foreseeable future (Hall *et al.* 2013). The UNWTO predicts the number of international tourist arrivals will increase by an average 3.3 per cent per year between 2010–30 (an average increase of 43 million arrivals a year), reaching an estimated 1.8 billion arrivals by 2030 (UNWTO 2011). It is fundamentally unclear how more sustainable absolute reductions in emissions and 'carbon neutral' positions are to be achieved given expected growth rates (Peeters and Landré 2011; Scott *et al.* 2012b; Hall *et al.* 2013).

Despite forecast increases in absolute emissions from tourism, the notion of tourism 'green growth', i.e. that you can continue to have economic growth and visitor growth at the same time as becoming more sustainable, has become an integral component of industry discourse on tourism and sustainability (UNWTO and UNEP 2011). Although the optimism of such a growth paradigm based on material/resource/energy efficiency is admirable, major changes in the energy mix to renewables and continued increases in visitor numbers appears problematic given arithmetic constraints of growth and efficiency limits, governance and market limits, and system limits (Hall 2009; Hoffmann 2011), as well as the general failure of considering rebound effects (Hall 2009, 2010a; Sorrell *et al.* 2009; Arvesen *et al.* 2011). It is therefore perhaps not surprising that some commentators take the position that de-carbonization of the economy and society will only be achieved if current consumption patterns, methods and lifestyles are also subject to profound change (Hoffmann 2011). Therefore the development of more ethical forms and patterns of ethical consumption would seem to be fundamental to improving the sustainability of tourism.

Ethical consumption in tourism can be understood as both an individual and collective process, given that individuals not only make individual decisions about their consumption and purchasing but may also be members of organizations and networks that promote forms of ethical consumption, e.g. slow food, farmers markets, fair trade organizations. Concerns with ethical consumption fits with a long interest in tourism of developing product characteristics and types that are more beneficial to destination communities and the environment, and which are often positioned as an alternative to 'mass tourism'. These include 'alternative tourism' (Tangi 1977; Dernoi 1981; Cohen 1987), 'community tourism' (Murphy 1983), 'green tourism' (Jones 1987), 'sustainable tourism' (Ap and Var 1990; Stevens 1990), 'ethical tourism' (Hall and Weiler 1992), 'responsible tourism' (Botterill 1991; Wheeller 1990), 'just tourism' (Hultsman 1995), 'metatourism' (Kariel and Draper 1990), 'sanfter tourismus' [gentle tourism]/'soft tourism'/'tourisme doux' (Danz 1985; Krippendorf 1987), 'appropriate tourism' (Singh *et al.* 1989; Hall 1994), 'ecotourism' (Hemmi 1982; Hall 1984) and the more recent 'slow tourism' (Matos 2004) and 'steady-state tourism' (Hall 2009). Although perhaps a number of the 'tourisms', if not all, have subsequently lost some of their conceptual power since first being developed, they do highlight that: (a) there is a long tradition of concern with ethical consumption in tourism, and (b) despite academic and even policy

Can ethical consumption be sustainable? 35

success, with respect to being noted in documentation and becoming part of tourism policy discourses, tourism is clearly less sustainable than ever.

What the different forms of tourism noted above also highlight is the extent to which there has been an attempt to develop an alternative economic framework for the production and consumption of tourism at a lower cost to the environment and with less community and cultural impact. As this chapter highlights, tourism is not alone in this and has mirrored developments in other areas that are usually grounded in notions of community, localism and/or the development of short supply chains, social justice, and the development of new notions of citizenship that are discussed in more detail below. The chapter then goes on to review the different ways in which consumer behaviour change is framed, and highlights the links between different modes and understandings of governance (and therefore state intervention), and assumptions with respect to how change may occur. The recognition of the potential implications of structure and systems of provision in framing change possibilities provides a significant counterpoint to issues of marketization and the neoliberalization of tourism. However, as the chapter highlights, the supposed primacy of the market and freedom to choose encourages commodity fetishism, including that with respect to supposedly ethical forms of tourism consumption. This therefore raises, as the chapter addresses in the final section, fundamental questions about the possibilities of alternative tourisms and consumptions within the contemporary neoliberal capitalist project.

Citizenship

Although citizenship is a politically contested and evolving term, the notions of 'food citizenship' and 'ecological citizenship' refers to the development of an environmentally informed morality that provides a 'rationale for changing behaviour towards more sustainable lifestyles' (Seyfang 2011: 58). In the case of tourism and hospitality this is expressed, for example, in a number of different food chains including the growth of farmers' markets; the development of short food chains from producer to consumer; the growth of organic, ethically produced and free-range food products; and fair trade (Cohen 2006; Parkins and Craig 2006; Hall 2010b, 2012, 2013a, 2013b), all of which have had an impact on notions of sustainability with respect to food and in the wider context (e.g. Gössling *et al.* 2011; Hall and Gössling 2013).

Although much of the focus of food citizenship is on the development of new culinary systems that rely on face-to-face or spatially proximate food chains (Hall and Gössling 2013), Sage (2007) also identifies the development of spatially extended short supply chains for Fair Trade products. Fair Trade is a social and economic movement with respect to the trading of commodities and services between the developed and developing nations, and is significant for the development of a number of ethical tourism and hospitality products (Gössling and Hall 2013). Originally primarily a European social justice movement that sought to ensure a fair and equitable price for products and services imported by the

West (often from former colonies), the notion of fair trade has since expanded to include concern for the environment as well as general principles of sustainable development (Fridell 2007; Grankvist *et al.* 2007; Hall 2010b).

Fair Trade was defined by FINE (the four major international groups associated with fair trade: International Fair Trade Association/International Federation for Alternative Trade (IFAT, now renamed the World Fair Trade Organization (WFTO)), Fair Trade Labelling Organizations International (FLO), Network of European Worldshops (NEWS!) and the European Fair Trade Association (EFTA), as

> a trading partnership, based on dialogue, transparency and respect, that seeks greater equity in international trade. It contributes to sustainable development by offering better trading conditions to, and securing the rights of, marginalized producers and workers – especially in the South.
>
> (FINE 2001)

Although there are a large number of internationally traded food and non-food Fair Trade products, coffee and chocolate are the main ones and are found in many different hospitality and food service businesses (Hall 2010b). In addition, fair trade has also affected notions of social justice and international trade in tourism services (Evans 1994; Cleverdon 2001) that influenced notions of pro-poor tourism (Hall 2007).

Alternative economies and politics

Alternative economies have been a focus of research in the social sciences for several decades, especially as part of counter-institutional research. Much of this has been grounded in both drawing attention to the role of capitalism as well as seeking to form alternatives and oppositional approaches to it (Britton 1992). The influential work of Gibson-Graham (1993, 1995, 1996, 2003a, 2003b, 2008) called for critics of capitalism to both recognize and enact alternatives to dominant capitalist formations. Gibson-Graham (1996) emphasized that discourses that assume capitalist subjects and hegemony, what was termed 'capitalonormativity', in fact blind us to the multiplicity of economic relationships already present around us. Thus, the search for and recognition of alterity is 'a vital political act' (McCarthy 2006: 804). Indeed, Carrier (2010: 672) notes, that 'ethical consumption marks a conjunction of capitalism and conservation, for it identifies people's market transactions, and market mechanisms generally, as the effective way to bring about protection of the environment'. Nevertheless, for some the suggestion that fairer and more sustainable alternatives are there if we are simply willing to open our eyes and see them is tantamount to suggesting that we respond to the depredations of global capitalism by 'donning rose-coloured glasses' (e.g. Watts 2003; Lovell 2009).

The potential of consumption to be a site of progressive politics or 'caring at a distance' is a topic of substantial debate (Leyshon *et al.* 2003; Lovell 2009;

Fuller *et al.* 2010). Many have argued that consumption can and should be interpreted as an important site of moral and other regarding behaviour (Sayer 2003; Newholm and Shaw 2007; Trentmann 2007; Seyfang 2011). Such notions have become an important part of contemporary public policy given that the range of policy measures that the state utilizes to achieve its policy goals are based on assumptions regarding individual and collective behaviour (Hall 2011b, 2011c). Although usually not explicitly stated, such assumptions are contained not only in public policy positions but also in the recommendations of industry, institutions, consultants and academics on how to improve the sustainability of tourism because they implicitly suggest that by engaging a specific policy setting 'A' consumers will do 'B'. Or, to be more precise 'C', given the dominance of the paradigm of 'ABC' – attitude, behaviour, and choice – in framing consumer behaviour (Shove 2010).

Framing consumer change

Three major contemporary approaches can be recognized in approaching issues of sustainability, tourism mobility, as well as other areas of sustainable consumption: utilitarian, social/psychological and systems of provision/institutional (Whitmarsh 2009; Seyfang 2011) (Table 3.1). The utilitarian approach to behavioural change utilizes a conventional neoclassical microeconomic view of consumption by individuals as rational utility-maximizers. The approach assumes that individuals consume goods and services in free markets with perfect competition and information in order to decide a course of action that delivers the greatest utility to the individual. From this 'consumer sovereignty' perspective, efforts to promote sustainable mobility and consumption tend to rely on government intervention (or self-regulation) to correct 'market failure' and ensure that private and corporate individuals have greater information on which to base their decisions.

The approach aims to appeal to rational actors with information to overcome an 'information deficit' and encourage 'rational behaviour' and send appropriate signals to the market. Interestingly, Newholm and Shaw (2007) note that, at its more extreme end, this could possibly also include consumer buycotts and boycotts as consumer citizens, although this is likely not the intention of public policies. Indeed, Barnett *et al.* (2005b: 45), in examining campaign group localities, argued that ethical consumption could 'enrol ordinary people in active political engagement. Roles of consumer and citizen would not necessarily, therefore, be exclusive'.

The failure of neoclassical economic models to increase levels of sustainable consumption has led to the realization that behaviour does not change simply because of better quality information (Whitmarsh 2009; Whitmarsh *et al.* 2009). The critique of neoclassical/rational models has primarily come from two social/psychological sources: behavioural economics and consumption studies. Behavioural economics recognizes that individuals have bounded rationality and often engage in satisficing behaviour as well as the role of social norms and routines

Table 3.1 Approaches to consumer change

Approach	Scale	Understanding of decision-making	Consumption is …	Tools to achieve sustainable consumption	Dominant forms of governance
Utilitarian (consumer sovereignty; green economics)	Individual	Cognitive information processing on basis of rational utility-maximization	The means for increasing utility	Green labelling, tax incentives, pricing (including carbon trading), education	Markets (marketization and privatization of state instruments)
Social and psychological (behavioural economics/ green consumerism/ABC model)	Individual	Response to psychological needs, behaviour and social contexts. Dominant paradigm of 'ABC': attitude, behaviour, and choice	Satisfier of psychological needs; cultural differentiator; marker of social meaning and identity	Nudging – making better choices through manipulating a consumer's environment. Social marketing in order to encourage behavioural change and promote sustainable lifestyles and behaviour	Markets (marketization and privatization of state instruments); Networks (public–private partnerships)
Systems of provision/ institutions (degrowth, steady-state tourism)	Community, society, network	Constrained/shaped by sociotechnical infrastructure and institutions	Routine habit, inconspicuous rather than conspicuous	Short supply chains, local food, local tourism	Hierarchies (nation state and supranational institutions); Communities (public–private partnerships, communities)

Sources: Hall 2009a, 2011b, 2011c, 2013a; Whitmarsh 2009; Shove 2010; Seyfang 2011; Gössling and Hall 2013.

that are not subject to rational cost–benefit calculations, including notions of community and fairness in economic outcomes. Significantly, with respect to the manner of influencing behaviour, it also stresses that too much choice in the market leads to information overload and subsequent difficulties in decision-making (Seyfang 2011).

A key policy focus in encouraging more sustainable consumer behaviour is the notion of 'nudging' (Cialdini 2007; Thaler and Sunstein 2008). The focus of nudging is in reconfiguring the 'choice architecture' to encourage beneficial decision-making by consumers. The approach suggests that the goal of public policy-making should be to steer citizens towards making positive decisions as individuals, and for society, while preserving individual choice. Acting as 'choice architects', policy makers organize the context, process and environment in which individuals make decisions and, in so doing, they exploit 'cognitive biases' to manipulate people's choices (Alemanno 2012). In the UK the influence of the concept on policy initiatives, including emissions reduction and changes in travel behaviour, is seen in the MINDSPACE report which although published during the tenure of the Cameron government had been commissioned by the previous government, thereby illustrating the approach's broad political attraction. MINDSPACE is a mnemonic 'which can be used as a quick checklist when making policy' (Dolan *et al.* 2010: 8),

M: Messenger we are heavily influenced by who communicates information
I: Incentives our responses to incentives are shaped by predictable mental short-cuts such as strongly avoiding losses
N: Norms we are strongly influenced by what others do
D: Defaults we 'go with the flow' of pre-set options
S: Salience our attention is drawn to what is novel and seems relevant to us
P: Priming our acts are often influenced by sub-conscious cues
A: Affect our emotional associations can powerfully shape our actions
C: Commitments we seek to be consistent with our public promises, and reciprocate acts
E: Ego we act in ways that make us feel better about ourselves

A key insight of such an approach is that people do not act as isolated individuals, and instead consumption is socially situated and is often deeply embedded, with habits and norms being significant. An insight that reflects that consumption is a multilayered phenomenon that is full of meaning including its role as a signifier of identity, cultural and social affiliations (consumption tribes), and relationships (Seyfang 2011).

Although social/psychological approaches recognize the need to make major behavioural changes, and while social norms are often cited as driving factors, they do not examine how needs and aspirations come to be as they are, i.e. they do not fundamentally question structures and paradigms (Shove 2010). As in neoclassical economics the focus is on consumer choice,

> relatively little attention is paid to the larger and longer-term processes by which consumers acquire their tastes or their utility functions, and by which the object of consumption comes into being and is brought to the presence of the consumer.
>
> (Carrier 2010: 680)

Although some research has examined how everyday consumption practices have altered in response to environmental problems (Southerton et al. 2004a), this work remains relatively isolated from approaches which continue to configure consumption as an individualized activity so that 'the drivers and mechanisms involved are seen to boil down to a matter of individual choice' (Southerton et al. 2004b: 3). Instead, the ways in which 'processes of consuming are configured by many aspects of production which have a structuring effect on what goods and services are provisioned, how those goods and services shape the consumption of related products, and how objects are used' are neglected (Southerton et al. 2004b: 7). Similarly, Lovell et al.'s (2009) research on carbon offsetting suggests that how the practice of consumption is thought about needs to reflect a broader sense of 'who' and 'what' consumption involves, and incorporate institutions and organizations as well as individuals as a subject, participant, or consumer, ecological or political citizen (Seyfang 2005, 2007).

Therefore a more structural perspective on the organization of systems of consumption and provision, focuses on the contextual collective societal institutions, norms, rules, structures and infrastructure that constrain individual decision-making, consumption and lifestyle practices. These are referred to variously as 'infrastructures of provision' (Southerton et al. 2004a), institutions, and 'systems of provision': the vertical commodity chains comprising production, finance, marketing, advertising, distribution, retail and consumption that 'entails a more comprehensive chain of activities between the two extremes of production and consumption, each link of which plays a potentially significant role in the social construction of the commodity both in its material and cultural aspects' (Fine and Leopold 1993: 33).

The significance of the systems of provision approach is that it emphasizes that particular sociotechnical systems constrain choice to that available within the system of provision, and can therefore 'lock in' consumers to particular ways of behaving and consuming (Seyfang 2011). The approach suggests that positioning the problem of sustainable tourism consumption as a problem of personal choice fails to appreciate the socially situated and structured nature of consumption and is an 'arguably suspect theory of choice' given that it assumes

> that people could and would act differently if only they knew what damage they were doing. Such ideas inform programmes of research into the relationship between environmental belief and action, and the design of policy initiatives geared around the provision of more and better information.
>
> (Southerton et al. 2004b: 4)

Much research on consumption fails to recognize that consumers do not consume resources but instead they consume the services that are made possible by resources. A focus on the end consumer also obscures 'important questions about the production and manufacturing of options and the intersection between, design, demand and use' (Southerton *et al.* 2004b: 5), while the social dimensions of consumption, such as distribution and equity effects, are ignored and conceptually 'closed off' (Southerton *et al.* 2004b: 5). Perhaps significantly with respect to the development of broader critiques, it suggests that sociotechnical systems, institutions and structures are not neutral, in that their formation and constitution is likely to sway behaviours more in one direction than others.

Ethical consumption and commodity fetishism

The recognition of the potential implications of structure and systems and provision runs counter to notions of the autonomous individual acting in a market free of constraint. This freedom to choose is generally seen as important for human welfare, as it allows each person to satisfy their wants and because, 'without it, people would develop distorted tastes and values; only the free, in other words, can come to recognise and value the good' (Carrier 2010: 674). Yet, as Carrier goes on to note,

> While some claim ... that market transactions by autonomous actors are natural, it is perhaps more accurate to refer to them as 'naturalised'. That is because the natural basis and positive value assigned to autonomy and market freedom reflect a view of people and the world around them that is peculiar to some political-economic systems rather than others. That view tends to fetishise commodities, market transactions and, indeed, people themselves.

The notion of commodity fetishism is extended here to include concern with the general tendency to obscure the people, processes and loss of natural capital that are part of creating a product, whether commodity or service, and of bringing it to market for commercial gain (Castree 2001; Carrier 2010, 2012). The notion of a commodity also refers to the commoditization of services, which is especially significant for some aspects of tourism as well as sustainability tools such as carbon offsets as they 'differ in important ways from fair trade and organic products because of their nontangible nature, the lack of direct material benefits, and the current paucity of regulation or certification' (Lovell *et al.* 2009: 2359). Nevertheless, regardless of the nature of the product, fetishism is identified as occurring in three interrelated ways of the object that is to be purchased, of the acquisition or consumption of that object, of the environment to be protected (Carrier 2010, 2012).

It is exceedingly difficult for alternative economies to maintain their differences from the global capitalist economy (Mutersbaugh 2005a, 2005b). In both food and fair trade literature, there is a clear consensus in both that the standardization and verification of certifications guaranteeing 'quality' of multiple sorts ensures premiums but also creates rents, imposes costs on producers, and excludes many potential beneficiaries. There is an equal consensus on the power of retailers.

(McCarthy 2006: 808)

Such processes lead to mainstream business dynamics in supposedly alternative networks as they operate in the same political-economic system, which constrains them to attract customers in the same way, by presenting appealing images that fetishize what they are selling so that they can effectively compete in a marketplace for a finite number of potential customers. Significantly, although alternative, they still have to sell with corresponding implications for other alternative networks and products. According to McCarthy (2006: 808) such networks and organizations promise an unveiling of the commodity fetish in that they deliberatively position themselves against 'mainstream' consumption as an alternative, while simultaneously constructing fetishes of their own. Indeed, according to Cafédirect, the UK's largest Fair Trade hot drinks firm, fair trade

> emerged out of social movements in both producer and consumer countries, but it is a market-based approach to development. *So difficult to support the anti-capitalism criticism.* It operates within the limits of international trade, but tries to create opportunities to make the trade more beneficial to those who are most disadvantaged, while also campaigning to try and make some change[s] to the current world trade system [emphasis added].
>
> (Cafédirect 2009)

The focus on profit by Cafédirect, and their competitive position in the market, was also emphasized in an interview with its head, Penny Newman: 'If you want to change the trading system you've got to be on the same terms as the conventional system. You need to make a profit; it's what you do with the profit' (in Martinson, 2007).

The above point has also been well made with respect to the current alliances between capitalism and conservation in relation to ecotourism. These alliances are often characterized by an 'aggressive faith' in market solutions to environmental problems (Brockington *et al.* 2008), including the development of nature-based tourism, that allows capitalism to identify, open and colonize new spaces in nature (Duffy and Moore 2010). 'They are actively remaking economies, landscapes, livelihoods and conservation policy and practice and they are partying in the symbolic heartlands of capitalism' (Brockington and Duffy 2010: 470).

The link between capitalism and conservation is displayed in the way that nature and conservation can provide an avenue for corporations to portray themselves as 'green' via sponsorship of particular species or conservation projects (Robinson 2012). This means that the commercialization of international

Can ethical consumption be sustainable? 43

biodiversity conservation is not only 'creating new symbolic and material spaces for global capital expansion' (Corson 2010: 578), but paradoxically is also fuelling the very processes of capital accumulation that have led to species and environments becoming endangered in the first place. In one sense there is nothing new in the relationship between capitalism and conservation – the first national parks were, after all, created because they were regarded as worthless for other commercial activities such as agriculture, forestry or mining but could assume an economic value on the basis of their aesthetic picturesque qualities for tourism (Hall and Frost 2009). But what is 'new' is the vast scope of the ecotourism project by which the tourism-derived economic values of particular ecosystem services are reified as justifications for biodiversity conservation and the extent to which this is regarded as being ethically based. This reflects the notion, conveyed by organizations such as UNWTO, that resolutions to environmental problems 'hinge on heightened commodity production and consumption, particularly of newly commodified ecosystem services' (Brockington and Duffy 2010: 472) as part of a green growth economy (UNWTO and UNEP 2011; Lipman *et al.* 2012).

Spectacle has long been acknowledged as critical to the imaging of tourism in an urban context, but is also important to nature-based tourism, where representations of, and connection to, places, people and causes has long been mediated through commodified images (Brockington and Duffy 2010; Igoe 2010). Igoe *et al.* (2010: 502) argue that in consuming these images tourists are given 'the romantic illusion that they are adventurously saving the world' while the deleterious ecological impacts of these very purchases, particularly their carbon emissions and ecological footprint, and the lifestyles they require, are ignored or 'neatly erased' (Brockington and Duffy, 2010: 472).

> By focusing consumers' attention on distant and exotic locales, the spectacular productions ... conceal the complex and proximate connections of people's daily lives to environmental problems, while suggesting that the solutions to environmental problems lay in the consumption of the kinds of commodities that helped produce them in the first place.
> (Igoe *et al.* 2010: 504)

Within the society of the spectacle, Carrier (2010) argues that ethical consumption depends upon the circulation of images that are taken to denote ethicality, and entails forms of fetishism that subvert ethical consumption's central goals (Brockington and Duffy 2010). However, the validity and ability of certain images to circulate as a common currency of ethical consumption requires a high degree of consumer ignorance in order to operate, i.e. the reality behind the image is not made clear.

Like Igoe *et al.* (2010), Carrier (2010) observes that ethical consumption often ignores the broader environmental, and socioeconomic contexts in which it operates. As Brockington and Duffy (2010) suggested, ethical consumption often achieves its ethically positive results by counting various aspects of the production and consumption of its commodities and not counting others. The structural constraints of systems of provision can clearly limit the good intentions of ethical consumption. However, Carrier (2010) identified a second, more insidious, effect at work. Underlying ethical consumption, and reinforced by every 'ethical' purchase, is the belief that 'personal consumption decisions ... are an appropriate and effective vehicle for correcting ... the ill effects of a system of capitalist production' (Carrier 2010: 683). Brockington and Duffy (2010) believe that this too decontextualizes the consumer, hiding the structures and institutions that encouraged them to seek solutions in consumption in the first place.

For example, while whale watching tours are usually positioned as an economically viable option to the commercial whaling industry, they typically fail to consider the ecological cost of tourists travelling to participate in such activities or the ecological disturbance that whale watching itself causes (Neves 2010; Mustika et al. 2012a, 2012b; Neves and Igoe 2012). Similarly, tourists to the world's polar regions, while portrayed as conservation ambassadors, often accrue enormous per capita emissions (Hall 2010b). On a per capita basis the 14.97 tonnes of GHG produced during the typical two-week travels of the Antarctic tourist is equal to the total emissions produced by an average European in 17 months (Amelung and Lamers 2007).

The disassociation from production that characterizes contemporary consumption in developed economies means that consumer satisfaction, and even the consumption experience, may be unencumbered by 'reality' (Borgmann 2000). This situation potentially allows consumers, freed from basic needs, to become more responsible for their consumption behaviour as 'economic voters' (Dickinson and Carsky 2005). Parkins and Craig (2006) draw on the notion of the 'risk society' in their assessment of the slow food movement, and suggest that in an individualized culture consumers increasingly face the consequences of their choices without the benefit of traditions to guide them (Newholm and Shaw 2007). Indeed, Barnett et al. (2005a: 23) argue that commodity consumption has been 'problematised' such that 'ethical consumption ... involves both a governing of consumption and a governing of the consuming self'. In such a situation organizations that establish green and sustainable products and schemes, such as in the case of tourism organic food certification, green hotels and carbon offset schemes, become critical nodes in the circuits of production and consumption (Lovell et al. 2009).

Goodman (2004: 898) argues that new fetishes can still be used strategically for progressive ends and writes of the development of 'discursive fields' around Fair Trade products as the 'second moment of production'; narratives are produced and then consumed. Bryant and Goodman (2004), however, represent many who argue that alternative commodity networks effectively enable the perpetuation of larger inequalities. Barnett et al. (2005a) question the assumptions regarding rationality and information underlying both sides of this debate, and emphasize the complex dynamics of distinction and self-governance always at play in consumption. Cherrier (2005), for example, regards ethical consumption more as a 'search for meanings in life' where those meanings are both partial and are being continually socially and economically (re)negotiated.

The difficulty is that commercial organizations that are seeking to sell green products to ethical consumers 'are alert to how consumption practices "produce" narratives (or want to), which in turn feeds back to influence production' (Lovell *et al.* 2009: 2376). The reflexive process creates a 'delicate interplay' between narratives, consumption and production. In the case of carbon offsetting, Lovell *et al.* (2009) noted that rather than production driving consumption, what we are witnessing in the making of markets is a more complex process in which the practices of consumption, orchestrated through commercial organizations selling green products, are having a material effect on the ways in which those products are produced. Yet they emphasized that this is not being driven by individualized consumers. Instead, 'making carbon offset consumption possible is in many ways taking place from the top down, as offset organisations, corporations, NGOs, and the state compete to define and provide "ethical" consumption in relation to climate change' (Lovell 2009: 2376).

Narratives and their associated technologies are critical here in connecting consumption and production, and in so doing create new consumption arenas and spaces (Lovell *et al.* 2009: 2375). Rather than ethical consumption creating an alternative economic space, new spaces of capitalism have been opened up in existing systems of provision. As Goodman (2004: 896) explains in the context of Fair Trade:

> The production of meanings in the consumption of fair trade foods is both material and semiotic in the imaginary of fair trade commodities, that, while involved in connecting the places of consumption and production, also makes place through morally-tinged markets, premiums and standards.

As West (2010: 710–1) notes, 'This supposed embedding of the political and social into capitalist consumption may make for a tasty cup of coffee but it makes for lukewarm political action.'

Is there an alternative tourism?

Questions of what precisely makes such projects 'alternative' are central to issues of sustainable tourism, as are a set of questions about the relationship between the alternative and its constituent other, the mainstream or conventional capitalist economy (Whatmore *et al.* 2003; Watts *et al.* 2005). Few alternative products are so alternative that they eschew the circulation of capital in commodity form altogether; rather, they attempt to harness intrinsic dynamics of capitalism to progressive political projects. 'Alternatives' that reinforce notions of consumer sovereignty, state incapacity, and reified communities may themselves be neoliberal in important respects and serve to reinforce the focus on market solutions, deregulation and reduced state intervention or ownership (Castree 2006). This is very much reflected in research that contests the idea that tourism offers a neat solution to conservation and environmental problems, and instead argues that it acts as a driver of capitalism, particularly the ways that tourism extends and deepens neoliberalism (Brockington and Duffy 2010). Given that capitalism is identified as part of the problem, it is therefore difficult, if not impossible, for it to offer the solution. Neoliberalism does not necessarily displace or obliterate existing ways of valuing, owning and approaching nature (Duffy and Moore 2010); instead it mixes with local context to create new dynamics (Brockington and Duffy 2010) and thereby reflects the fundamentally uneven nature of the neoliberal project (Peck and Tickell 2002; Harvey 2005; Castree 2008a, 2008b).

Nevertheless, the way that tourism acts to provide an economic value for nature is integral to the way that neoliberalism is fundamentally concerned with 'the financialisation of everything' (Harvey 2005: 33). The problem with the rhetoric of valuing nature and tourism's contribution to biodiversity conservation is that it comes across, at least in industry material, as being such a positive form of ethical consumption. 'It presents us only with market solutions, win–win solutions (or win–win–win and more), ethically traded commodities, saved nature, wholesome communities, integrated landscapes, sustainable development, cleansed reputations and secure conservation brands' (Brockington and Duffy 2010: 481). 'It has been part of the genius of neoliberal theory to provide a benevolent mask full of wonderful-sounding words like freedom, liberty, choice and rights to hide the grim realities of the restoration or reconstitution of naked class power' (Harvey 2005: 119).

Yet, as suggested above, a closer examination of ethical tourism consumption versus the actual reality of the social, political and environmental implications of the supply and value chains that lead to the consumption experience, may provide an opening for revealing the often huge gap between promise and action. In such contested spaces some commentators believe that it may be possible to actively foster non-capitalist alternatives (e.g. Sayer 2003; Newholm and Shaw 2007; Seyfang 2011). For example, Sen and Majumder (2011) argue that fair trade constitutes a contested moral terrain that mediates between the visions of justice harboured by producers and activists in the Global South and the reflexive practices of Western consumers that leaves open the potential for creative iterations of the Fair Trade idea to give voice to the situated struggles of producer communities for justice.

There are certainly attempts by consumers to escape the 'totalizing logic of the market' by attempting to construct localized 'emancipated spaces' that are constructed by 'engaging in improbable behaviors, contingencies, and discontinuities' (Firat and Venkatesh 1995: 255). Such improbable behaviours include distinguishing acts that appear to be outside the logic of commercialization, such as voluntary simplicity and anti-consumption. These are all part of the new politics of consumption (Schor 2003). Nevertheless, focusing on consumption behaviour and lifestyle alone, without challenging the role of structure and the cultural forces of production, may well mean that alternative consumption paths become 'appropriated by experts, packaged, and sold, [and] loses its distinctive character. When this happens, even a lifestyle based on anticonsumption becomes defined in terms of commodities, possessions, sign value, and commercial success' (Murray 2002: 439).

Several themes therefore emerge in considering ethical consumption in tourism. While desires for environmental quality may foster the production of, what appear at first glance, alternatives to conventional hospitality and tourism products, inasmuch as the attendant dynamics are still focused on satisfying the self-interests of rational individual consumers within the contemporary capitalist economy, they are not 'alternative' in any deep sense (McCarthy 2006). Furthermore, the underlying assumption that is played out in behavioural economics, consumer behaviour and social marketing that consumption is

> about the maximization of individual utility in one way or another (albeit played out through complex dynamics of distinction and identity) runs the risk of perpetuating a long and mistaken refusal to take consumption seriously as a theoretical and political moment in the circulation of capital, and effectively denies the possibility of reflexivity.
>
> (McCarthy 2006: 809)

Indeed, in criticizing the economistic approach to consumption, Hansen and Schrader (1997: 443) state: 'In view of the reality of modern societies, it is neither possible nor ethically justifiable to make purchase decisions according to the individual maximisation of utility only.' Nevertheless, this perspective, and the ABC model that accompanies it, is the dominant paradigm in contemporary thinking on how sustainability can be achieved (Shove 2010).

There is also a need in tourism for a clearer theorization of what is alternative or not (Whatmore *et al.* 2003). Although this is a long-held point of debate in tourism studies (e.g. Cohen 1987) such a discussion should connect to some of the broader debates on alterity and alternative economic and political spaces (Fuller *et al.* 2010). Such theorizations need to engage more substantially to the embedding of tourism in contemporary neoliberal capitalism and its implications (Britton 1991; Fletcher 2011; Hall 2011a). Indeed, a further point that reflects McCarthy's (2006) on rural alternative economies in general, is that despite the promotion of Fair Trade, ecotourism and pro-poor tourism as alternative tourism, such economies have made terribly little progress in changing deeply entrenched North–South dynamics. 'Power in them remains overwhelmingly in the hands of consumers, certifiers, and retailers in the North' (McCArthy 2006: 809).

As Duffy (2012) noted with respect to the neoliberalization of nature, global ecotourism has targeted and opened up new frontiers in nature, which serves to expand and deepen neoliberalism with ecotourism acting as an environmental fix for capitalism (Fletcher and Neves 2012) that redraws the boundaries of access to nature as well as the distribution of costs and benefits. As Freidberg (2003b: 98) observed: '[W]e still need to consider how emerging networks, for all their newness, may in fact be laid down in the deep ruts worn by earlier relations of domination and extraction.'

So can tourism be sustainable? The answer is not in the foreseeable future. This is not to suggest that ethical consumption is without value for tourism. Ecological and consumer citizenship, including buycotts and boycotts, can certainly influence the behaviours of individual companies and agencies and provide some changes to policy instruments. Yet such change is also incremental and does not fundamentally challenge the dominant paradigm of marketization and the consequent 'lock-in' of tourism to a particular sociotechnical system of provision that is only making tourism, along with much contemporary consumption, less sustainable than ever. Green growth is only more of the same, and yet is also another alternative tourism fetish that reinforces marketization and the obsession with economic and visitor growth that has made tourism unsustainable in the first place. The ethical consumer is still a consumer. If the ethical value of 'individual choice' leads to yet more emissions and worsening environmental change then how ethical is it? Instead, nudging and social marketing must be used together with a fundamental change to the sociotechnical system itself if tourism is to become sustainable. This is the most significant ethical challenge of all.

Activity

THINK
1 How the fuck can just about every tourism destination and major tourism body in the world, such as the UNWTO, actively seek to increase the number of tourists each year, and yet still tell us that tourism can be sustainable?

DO
2 Ask: Do such organizations really believe the crap they come out with? Or are they lying?

REFLECT
3 How ethical can I really be in my tourism career?

ACT
4 Do something. Protest. Occupy. Change the courses they teach at your university (you are now meant to be a consumer after all). Engage in guerrilla gardening. Promote Fair Trade. Don't travel somewhere to change the world. Change the value chains that bind, and the tourism system from within. Most of all: don't sell out. Do something.

Suggested reading

Brockington, D., Duffy, R. and Igoe, J. (2008) *Nature Unbound. Conservation, Capitalism and the Future of Protected Areas*, London: Earthscan.

Hall, C.M. (2011) 'Consumerism, tourism and voluntary simplicity: We all have to consume, but do we really have to travel so much to be happy?' in *Tourism Recreation Research*, 36: 298–303.

Hall, C.M. (2014) *Tourism and Social Marketing*. London: Routledge.

Acknowledgement

The discussion on the three different approaches to consumer behaviour and implications for sustainable tourism was initially presented at the workshop 'Psychological and Behavioural Approaches to Understanding and Governing Sustainable Tourism Mobility', Freiburg, Germany, 3–6 July 2012, and 'Tourism as Commodity Fetishism' to the 21st Nordic Symposium in Tourism and Hospitality Research in Umeå, Sweden, November 2012. The author is grateful for comments received as well as funding support from the Freiburg Institute for Advanced Studies and the University of Eastern Finland.

References

Alemanno, A. (2012) 'Nudging Smokers – The Behavioural Turn Of Tobacco Risk Regulation' in *European Journal of Risk Regulation*, 1/2012: 32–42.
Amelung, B. and Lamers, M. (2007) 'Estimating The Greenhouse Gas Emissions From Antarctic Tourism' in *Tourism in Marine Environments*, 4: 121–33.
Ap, J. and Var, T. (1990) 'Does Tourism Promote World Peace?' in *Tourism Management*, 11: 267–73.
Arvesen, A., Bright, R.M. and Hertwich, E.G. (2011) 'Considering Only First-Order Effects? How Simplifications Lead To Unrealistic Technology Optimism In Climate Change Mitigation' in *Energy Policy*, 39: 7448–54.
Barnett, C., Cloke, P., Clarke, N. and Malpass, A. (2005a) 'Consuming Ethics: Articulating The Subjects And Spaces Of Ethical Consumption' in *Antipode*, 37: 23–45.
Barnett, C., Cloke P., Clarke, N. and Malpass A. (2005b) 'The Political Ethics Of Consumerism' in *Consumer Policy Review*, 15 (2): 45–51.
Borgmann, A. (2000) 'The Moral Complexion Of Consumption' in *Journal of Consumer Research*, 26: 418–22.
Botterill, T.D. (1991) 'A New Social Movement: Tourism Concern, The First Two Years' in *Leisure Studies*, 10: 203–17.
Britton, S. (1991) 'Tourism, Capital, And Place: Towards A Critical Geography Of Tourism' in *Environment and Planning D: Society and Space*, 9: 451–78.
Brockington, D. and Duffy, R. (2010) 'Capitalism And Conservation: The Production And Reproduction Of Biodiversity Conservation' in *Antipode*, 42: 469–84.
Brockington, D., Duffy, R. and Igoe, J. (2008) *Nature Unbound. Conservation, Capitalism and the Future of Protected Areas*, London: Earthscan.
Bryant, R.L. and Goodman, M.K. (2004) 'Consuming Narratives: The Political Ecology Of "Alternative' Consumption"' in *Transactions of the Institute of British Geographers NS*, 29: 344–66.
Cafédirect (2009) 'You Ask, They Answer: Cafedirect', The Guardian Greenliving Blog, 14 July 2009. Online, available at: www.guardian.co.uk/environment/2009/jul/13/you-ask-cafedirect?INTCMP=SRCH (accessed 1 January 2013).
Carrier, J.G. (2010) 'Protecting The Environment The Natural Way: Ethical Consumption And Commodity Fetishism' in *Antipode* 42: 672–89.
Carrier, J.G. (2012) 'Dollars Making Sense: Understanding Nature In Capitalism' in *Environment and Society: Advances in Research*, 3 (1): 5–18.
Castree, N. (2001) 'Commodity Fetishism, Geographical Imaginations And Imaginative Geographies' in *Environment and Planning A*, 33: 1519–25.

Castree, N. (2006) 'From Neoliberalism To Neoliberalisation: Consolations, Confusions, And Necessary Illusions' in *Environment and Planning A*, 38: 1–6.

Castree, N. (2008a) 'Neo-Liberalising Nature I: The Logics Of De- And Re-Regulation' in *Environment and Planning A*, 40: 131–52.

Castree, N. (2008b) 'Neo-Liberalising Nature II: Processes, Outcomes And Effects' in *Environment and Planning A*, 40: 153–73.

Cherrier, H. (2005) 'Using Existential Phenomenological Interviewing To Explore Meanings Of Consumption', in R. Harrison, T. Newholm and D. Shaw (eds), *The Ethical Consumer*, London: Sage, 125–135.

Cleverdon, R. (2001) 'Introduction: Fair Trade In Tourism – Applications And Experience' in *International Journal of Tourism Research*, 3: 347–9.

Cohen, E. (1987) 'Alternative Tourism: A Critique' in *Tourism Recreation Research*, 12(2): 13–18.

Cohen, M.J. (2006) 'Sustainable Consumption Research As Democratic Expertise' in *Journal of Consumer Policy*, 29: 67–77.

Connolly, J. and Prothero, A. (2003) 'Sustainable Consumption: Consumption, Communities And Consumption Discourse' in *Consumption Markets and Culture*, 6: 275–91.

Cooper-Martin, E. and Holbrook, M.B. (1993) 'Ethical Consumption Experiences And Ethical Space' in *Advances in Consumer Research*, 20: 113–18.

Corson, C. (2010) 'Shifting Environmental Governance In A Neoliberal World: US AID For Conservation' in *Antipode*, 42: 576–602.

Danz, W. (1985) 'Sanfter Tourismus – Schlagwort Oder Chance? [Gentle Tourism – Just A Catchphrase Or A Real Prospect?]' in *Tourism Review*, 40 (2): 10–12.

Dernoi, L.A. (1981) 'Alternative Tourism: Towards A New Style In North-South Relations' in *International Journal of Tourism Management*, 2: 253–64.

Dickinson, R. and Carsky, M. (2005) 'The Consumer As Economic Voter' in R. Harrison, T. Newholm and D. Shaw (eds), *The Ethical Consumer*, London: Sage: 25–36.

Dolan, P., Hallsworth, M., Halpern, D., King, D. and Vlaev, I. (2010) *MINDSPACE: Influencing Behaviour Through Public Policy*, London: Cabinet Office and Institute for Government.

Duffy, R. (2012) 'The International Political Economy Of Tourism And The Neoliberalisation Of Nature: Challenges Posed By Selling Close Interactions With Animals' in *Review of International Political Economy*, DOI:10.1080/09692290.2012.654443.

Duffy, R. and Moore, L. (2010) 'Neoliberalising Nature? Elephant-Back Tourism In Thailand And Botswana' in *Antipode*, 42: 742–66.

Etzioni, A. (2003) 'Introduction: Voluntary Simplicity–Psychological Implications, Societal Consequences' in D. Doherty and A. Etzioni (eds), *Voluntary Simplicity: Responding To Consumer Culture*, Oxford: Rowan and Littlefield Publishers: 1–25.

Evans, G. (1994) 'Fair Trade: Crafts Production And Cultural Tourism In The Third World', in A.V. Seaton (ed.), *Tourism: The State of the Art*, Chichester: John Wiley and Sons: 783–91.

Firat, A.F. and Venkatesh, A. (1995) 'Liberatory Postmodernism And The Reenchantment Of Consumption' in *Journal of Consumer Research*, 22: 239–67.

FINE [IFAT (International Fair Trade Association), FLO (Fair Trade Labelling Organisations International), NEWS! (Network of European Worldshops) and EFTA (European Fair Trade Association)] (2001) Fair trade definition and principles as agreed by FINE in December 2001, European Observatory on Fair Trade Public Procurement, European Fair Trade Association. Online, available at: www.european-fair-trade-association.org/observatory/index.php/en/fairtrade (accessed 1 April 2012).

Fine, B. and Leopold, E. (1993) *The World of Consumption*, London: Routledge.
Fletcher, R. (2011) 'Sustaining Tourism, Sustaining Capitalism? The Tourism Industry's Role In Global Capitalist Expansion' in *Tourism Geographies*, 13: 443–61.
Fletcher, R. and Neves, K. (2012) 'Contradictions In Tourism: The Promise And Pitfalls Of Ecotourism As A Manifold Capitalist Fix' in *Environment and Society: Advances in Research*, 3 (1): 60–77.
Freidberg, S. (2003a) 'Cleaning Up Down South: Supermarkets, Ethical Trade And African Horticulture' in *Social and Cultural Geography*, 4: 27–43.
Freidberg, S. (2003b) 'Culture, Conventions And Colonial Constructs Of Rurality In South-North Horticultural Trades' in *Journal of Rural Studies*, 19: 97–109.
Fridell, G. (2007) *Fair Trade Coffee: The Prospects and Pitfalls of Market-Driven Social Justice*, Toronto: University of Toronto Press.
Fuller, D., Jonas, A.E. and Lee, R. (eds), (2010) *Interrogating Alterity: Alternative Economic and Political Spaces*, Farnham: Ashgate.
Gibson-Graham, J.K. (1993) 'Waiting For The Revolution, Or How To Smash Capitalism While Working At Home In Your Spare Time' in *Rethinking Marxism*, 6(2): 10–24.
Gibson-Graham, J.K. (1995) 'Identity And Economic Plurality: Rethinking Capitalism And "Capitalist Hegemony"' in *Environment and Planning D: Society and Space*, 13: 275–82.
Gibson-Graham, J.K. (1996) *The End of Capitalism (As We Knew It): A Feminist Critique Of Political Economy*, Oxford: Blackwell.
Gibson-Graham, J.K. (2003a) 'An Ethics Of The Local' in *Rethinking Marxism*, 15 (1): 49–74.
Gibson-Graham, J.K. (2003b) 'Enabling Ethical Economies: Cooperativism And Class' in *Critical Sociology*, 29: 123–61.
Gibson-Graham, J.K. (2008) 'Diverse Economies: Performative Practices For "Other Worlds"' in *Progress in Human Geography*, 32: 613–32.
Goodman, M.K. (2004) 'Reading Fair Trade: Political Ecological Imaginary And The Moral Economy Of Fair Trade Foods' in *Political Geography*, 23: 891–915.
Grankvist, G., Lekedal, H. and Marmendal, M., (2007) 'Values And Eco-And Fair-Trade Labelled Products' in *British Food Journal*, 109: 169–81.
Gössling, S. and Hall, C.M. (2013) 'Sustainable Culinary Systems: An Introduction' in C.M. Hall and S. Gössling (eds), *Sustainable Culinary Systems: Local Foods, Innovation, and Tourism & Hospitality*, London: Routledge, 1–44.
Gössling, S., Garrod, B., Aall, C., Hille, J. and Peeters, P. (2011) 'Food Management In Tourism. Reducing Tourism's Carbon "Foodprint"' in *Tourism Management*, 32: 534–43.
Hall, C.M. (1994) 'Ecotourism In Australia, New Zealand And The South Pacific: Appropriate Tourism Or A New Form Of Ecological Imperialism?' in E. Cater and G. Lowman (eds), *Ecotourism: a sustainable option?* Chichester: John Wiley, 111–36.
Hall, C.M. (2006) 'Travel And Journeying On The Sea Of Faith: Perspectives From Religious Humanism' in D. Timothy and D. Olsen (eds), *Tourism, Religion and Spiritual Journeys*, London: Routledge, 64–77.
Hall, C.M. (2007) 'Pro-Poor Tourism: Do "Tourism Exchanges Benefit Primarily The Countries Of The South"?' *Current Issues in Tourism*, 10: 111–18.
Hall, C.M. (2009) 'Degrowing Tourism: Décroissance, Sustainable Consumption And Steady-State Tourism' in *Anatolia: An International Journal of Tourism and Hospitality Research*, 20: 46–61.
Hall, C.M. (2010a) 'Blending Fair Trade Coffee And Hospitality' in L. Joliffe (ed.), *Coffee Culture, Destinations and Tourism*, Clevedon: Channel View Publications, 159–71.

Hall, C.M. (2010b) 'Tourism And Environmental Change In Polar Regions: Impacts, Climate Change And Biological Invasion' in C.M. Hall and J. Saarinen (eds), *Tourism and Change in Polar Regions: Climate, Environments and Experiences*, London: Routledge, 42–70.

Hall, C.M. (2011a) 'Consumerism, Tourism And Voluntary Simplicity: We All Have To Consume, But Do We Really Have To Travel So Much To Be Happy?' in *Tourism Recreation Research*, 36: 298–303.

Hall, C.M. (2011b) 'Policy Learning And Policy Failure In Sustainable Tourism Governance: From First And Second To Third Order Change?' in *Journal of Sustainable Tourism*, 19: 649–71.

Hall, C.M. (2011c) 'A Typology Of Governance And Its Implications For Tourism Policy Analysis' in *Journal of Sustainable Tourism*, 19: 437–57.

Hall, C.M. (2012) 'The Contradictions And Paradoxes Of Slow Food: Environmental Change, Sustainability And The Conservation Of Taste' in S. Fullagar, K. Markwell and E. Wilson (eds), *Slow Tourism: Experiences and Mobilities*, Clevedon: Channel View Publications: 58–68.

Hall, C.M. (2013a) 'The Local In Farmers Markets In New Zealand' in C.M. Hall and S. Gössling (eds), *Sustainable Culinary Systems: Local Foods, Innovation, and Tourism & Hospitality*, London: Routledge: 99–121.

Hall, C.M. (2013b) 'Why Forage When You Don't Have To? Personal And Cultural Meaning In Recreational Foraging: A New Zealand Study' in *Journal of Heritage Tourism*, DOI: 10.1080/1743873X.2013.767809.

Hall, C.M. and Frost, W. (2009) 'National Parks And The "Worthless Lands Hypothesis" Revisited' in W. Frost and C.M. Hall (eds), *Tourism and National Parks: International Perspectives On Development, Histories And Change*, London: Routledge, 45–62.

Hall, C.M. and Gossling, S. (eds) (2013) *Sustainable Culinary Systems: Local Foods, Innovation, and Tourism and Hospitality*, London: Routledge.

Hall, C.M. and Weiler, B. (1992) 'Introduction. What's Special About Special Interest Tourism?' in B. Weiler and C.M. Hall (eds), *Special Interest Tourism*, London: Belhaven Press: 1–14.

Hall, C.M., Scott, D. and Gössling, S. (2013) 'The Primacy Of Climate Change For Sustainable International Tourism' in *Sustainable Development*, 21 (2), DOI: 10.1002/sd.1562.

Hall, D. (1984) 'Conservation By Ecotourism' in *New Scientist* 1399: 38–9.

Hansen, U. and Schrader, U. (1997) 'A Modern Model Of Consumption For A Sustainable Society' in *Journal of Consumer Policy*, 20: 443–69.

Harvey, D. (2005) *A Brief History of Neoliberalism*, Oxford: Oxford University Press.

Hemmi, J. (1982) *Ecological Tourism (Ecotourism) – A New Viewpoint*. Report presented at FAO/ECE Working Party on Agrarian Structure and Farm Rationalization, Symposium on Agriculture and Tourism, Mariehamn, Finland, 7–12 June 1982.

Hoffmann U. (2011) *Some Reflections On Climate Change, Green Growth Illusions And Development Space*, UNCTAD Discussion Paper No. 205, Geneva: UNCTAD.

Hultsman, J. (1995) 'Just Tourism: An Ethical Framework' in *Annals of Tourism Research*, 22: 553–67.

Igoe, J. (2010) 'The Spectacle Of Nature In The Global Economy Of Appearances: Anthropological Engagements With The Spectacular Mediations Of Transnational Conservation' in *Critique of Anthropology*, 30: 375–97.

Igoe, J., Neves, K. and Brockington, D. (2010) 'A Spectacular Eco-Tour Around The Historic Bloc: Theorising The Convergence Of Biodiversity Conservation And Capitalist Expansion' in *Antipode*, 42: 482–512.

Jones, A. (1987) 'Green Tourism' in *Tourism Management*, 8 (4), 354–6.
Kariel, H.G. and Draper, D. (1990) 'Metatourism: Dealing Critically With The Future Of Tourism Environments' in *Journal of Cultural Geography*, 11: 139–55.
Krippendorf, J. (1987) *The Holiday Makers: Understanding The Impact Of Leisure And Travel*, London: Heinemann.
Leyshon, A., Lee, R. and Williams, C.C. (eds) (2003) *Alternative Economic Spaces*, London: Sage.
Lipman, G., DeLacy, T., Vorster, S., Hawkins, R. and Jiang, M. (eds) (2012) *Green Growth and Travelism – Letters from Leaders*, Oxford: Goodfellows.
Lovell, H., Bulkeley, H. and Liverman, D. (2009) 'Carbon Offsetting: Sustaining Consumption?' in *Environment and Planning A*, 41: 2357–79.
Martinson, J. (2007) 'Interview: Penny Newman. The Ethical Coffee Chief Turning A Fair Profit' in *The Guardian*, Friday 9 March 2007. Online, available at: www.guardian.co.uk/environment/2007/mar/09/food.business? (accessed 1 January 2013).
Matos, R. (2004) 'Can Slow Tourism Bring New Life To Alpine Regions?' in K. Weiermair and C. Mathies (eds), *The Tourism and Leisure Industry: Shaping the Future*, Binghampton: Howarth Press: 93–103.
McCarthy, J. (2006) 'Rural Geography: Alternative Rural Economies – The Search For Alterity In Forests, Fisheries, Food, And Fair Trade' in *Progress in Human Geography*, 30: 803–11.
Mohr, L.A., Webb, D.J. and Harris, K.E. (2001) 'Do Consumers Expect Companies To Be Socially Responsible? The Impact Of Corporate Social Responsibility On Buying Behaviour' in *Journal of Consumer Affairs*, 35: 45–72.
Murray, J.B. (2002) 'The Politics Of Consumption: A Re-Inquiry On Thompson And Haytko's (1997) "Speaking of Fashion"' in *Journal of Consumer Research*, 29: 427–40.
Murphy, P.E. (1983) 'Tourism As A Community Industry: An Ecological Model Of Tourism Development' in *Tourism Management*, 4: 180–93.
Musgrove, F. (1974) *Ecstasy and Holiness: Counterculture and the Open Society*, Bloomington: Indiana University Press.
Mustika, P.L.K., Birtles, A., Everingham, Y. and Marsh, H. (2012a) 'The Human Dimensions Of Wildlife Tourism In A Developing Country: Watching Spinner Dolphins At Lovina, Bali, Indonesia' in *Journal of Sustainable Tourism*, DOI:10.1080/09669582.2012.692881.
Mustika, P.L.K., Birtles, A., Welters, R. and Marsh, H. (2012b) 'The Economic Influence Of Community-Based Dolphin Watching On A Local Economy In A Developing Country: Implications For Conservation' in *Ecological Economics*, 79: 11–20.
Mutersbaugh, T. (2005a) 'Fighting Standards With Standards: Harmonization, Rents, And Social Accountability In Certified Agrofood Networks' in *Environment and Planning A*, 37: 2033–51.
Mutersbaugh, T. (2005b) 'Just-In-Space: Certified Rural Products, Labour Of Quality, And Regulatory Spaces' in *Journal of Rural Studies*, 21: 389–402.
Neves, K. (2010) 'Cashing In On Cetourism: A Critical Ecological Engagement With Dominant E-NGO Discourses On Whaling, Cetacean Conservation, And Whale Watching', *Antipode*, 42: 719–741.
Neves, K. and Igoe, J. (2012) 'Uneven Development And Accumulation By Dispossession In Nature Conservation: Comparing Recent Trends In The Azores And Tanzania' in *Tijdschrift Voor Economische En Sociale Geografie*, 103 (2), 164–79.
Newholm, T. and Shaw, D. (2007) 'Studying The Ethical Consumer: A Review Of Research' in *Journal of Consumer Behaviour*, 6: 253–70.

Parkins, W. and Craig, G. (2006) *Slow Living*, Oxford: Berg.
Peattie, K. and Collins, A. (2009) 'Guest Editorial: Perspectives On Sustainable Consumption' in *International Journal of Consumer Studies*, 33 (2): 107–12.
Peck, J. and Tickell, A. (2002) 'Neoliberalizing Space' in *Antipode*, 34: 380–404.
Peeters, P. and Landré, M. (2011) 'The Emerging Global Tourism Geography – An Environmental Sustainability Perspective' in *Sustainability*, 4 (1): 42–71.
Roberts, J.A. (1993) 'Sex Differences In Socially Responsible Consumers' Behaviour' in *Psychological Reports*, 73: 139–48.
Robinson, J.G. (2012) 'Common And Conflicting Interests In The Engagements Between Conservation Organizations And Corporations' in *Conservation Biology*, 26: 967–77.
Sage, C. (2007) 'Trust In Markets: Economies Of Regard And Spaces Of Contestation In Alternative Food Networks' in J. Cross and A. Morales (eds), *Street Entrepreneurs: People, Place and Politics In Local and Global Perspective*, Abingdon: Routledge: 147–63.
Sage, C. (2012) *Environment and Food*, London: Routledge.
Sayer, A. (2003) '(De)Commodification, Consumer Culture, And Moral Economy' in *Environment and Planning D: Society and Space*, 21: 341–57.
Schor, J. (2003) 'The Problem Of Over-Consumption – Why Economists Don't Get It' in D. Doherty and A. Etzioni (eds), *Voluntary Simplicity: Responding To Consumer Culture*, Oxford: Rowan and Littlefield: 65–82.
Scott, D, Gössling, S. and Hall, C.M. (2012a) 'International Tourism And Climate Change' in *WIRES Climate Change*, 3 (3), DOI: 10.1002/wcc.165.
Scott, D, Gössling, S. and Hall, C.M. (2012b). *Tourism and Climate Change: Impacts, Adaptation And Mitigation*, Abingdon: Routledge.
Sen, D. and Majumder, S. (2011) 'Fair Trade And Fair Trade Certification Of Food And Agricultural Commodities: Promises, Pitfalls, And Possibilities' in *Environment and Society: Advances in Research*, 2 (1): 29–47.
Sen, S., Grhan-Canli, Z. and Morwitz, V. (2001) 'Withholding Consumption: A Social Dilemma Perspective On Consumer Boycotts' in *Journal of Consumer Research*, 28: 399–418.
Seyfang, G. (2005) 'Shopping For Sustainability: Can Sustainable Consumption Promote Ecological Citizenship?' in *Environmental Politics*, 14: 290–306.
Seyfang, G. (2007) 'Cultivating Carrots And Community: Local Organic Food And Sustainable Consumption' in *Environmental Values*, 16: 105–23.
Seyfang, G. (2011) *The New Economics of Sustainable Consumption*, London: Palgrave Macmillan.
Shaw, D. and Newholm, T. (2002) 'Voluntary Simplicity And The Ethics Of Consumption' in *Psychology and Marketing*, 19: 167–85.
Shi, D.E. (1986) *The Simple Life: Plain Living And High Thinking In American Culture*, New York: Oxford University Press.
Shove, E. (2010) 'Beyond The ABC: Climate Change Policy And Theories Of Social Change' in *Environment and Planning A*, 42: 1273–85.
Singh, T.V., Theuns, H.L. and Go, F.M. (eds) (1989) *Towards Appropriate Tourism: The Case Of Developing Countries*, Frankfurt: Peter Lang.
Sorrell, S., Dimitropoulos, J. and Sommerville, M. (2009) 'Empirical Estimates Of The Direct Rebound Effect: A Review' in *Energy Policy*, 37: 1356–71.
Southerton, D., Chappells, H. and Van Vliet, B. (eds) (2004a) *Sustainable Consumption: The Implications Of Changing Infrastructures Of Provision*, Cheltenham: Edward Elgar.

Southerton, D., Van Vliet, B. and Chappells, H. (2004b) 'Introduction: Consumption, Infrastructures And Environmental Sustainability' in D. Southerton, H. Chappells and B. Van Vliet (eds), *Sustainable Consumption: The Implications Of Changing Infrastructures Of Provision*, Cheltenham: Edward Elgar: 1–14.

Stevens, T. (1990) 'Greener Than Green' in *Leisure Management*, 10 (9), 64–6.

Syed Marzuki, S.Z., Hall, C.M. and Ballantine, P.W. (2012) 'Restaurant Manager's Perspectives On Halal Certification' in *Journal of Islamic Marketing*, 3 (1), 47–58.

Tangi, M. (1977) 'Tourism And The Environment' in *Ambio*, 6 (6): 336–41.

Thaler, R.H. and Sunstein, C.R. (2008) *Nudge: Improving Decisions About Health, Wealth And Happiness*, London: Yale University Press.

Trentmann, F. (2007) 'Citizenship And Consumption' in *Journal of Consumer Culture*, 7: 147–58.

UNWTO (2011) *Tourism Towards 2030: Global Overview*. UNWTO General Assembly, 19th Session, Gyeongju, Republic of Korea, 10 October 2011. Madrid: United Nations World Tourism Organization.

UNWTO and UNEP (2011) 'Tourism: Investing In The Green Economy' in *Towards a Green Economy*, Geneva: United Nations Environmental Programme.

Vitell, S.J. (2003) 'Consumer Ethics Research: Review, Synthesis And Suggestions For The Future', *Journal of Business Ethics* 43: 33–47.

Watts, D.C.H., Ilbery, B. and Maye, D. (2005) 'Making Reconnections In Agro-Food Geography: Alternative Systems Of Food Provision' in *Progress in Human Geography*, 29: 22–40.

Watts, M. (2003) 'Development And Governmentality' in *Singapore Journal of Tropical Geography*, 24: 6–34.

Webster Jr., F.E. (1975) 'Determining The Characteristics Of The Socially Conscious Consumer' in *Journal of Consumer Research*, 2: 188–96.

West, P. (2010) 'Making The Market: Specialty Coffee, Generational Pitches, And Papua New Guinea' in *Antipode*, 42: 690–718.

Whatmore, S., Stassart, P. and Renting, H. (2003) 'What's Alternative About Alternative Food Networks?' in *Environment and Planning A*, 35: 389–91.

Wheeller, B. (1990) 'Responsible Tourism' in *Tourism Management*, 11: 262–3.

Whitmarsh, L. (2009) 'Behavioural Responses To Climate Change: Asymmetry Of Intentions And Impacts' in *Journal of Environmental Psychology*, 29: 13–23.

Whitmarsh, L., O'Neill, S., Seyfang, G. and Lorenzoni, I. (2009) *Carbon Capability: What Does It Mean, How Prevalent Is It, And How Can We Promote It?* Tyndall Working Paper 132, Tyndall Centre for Climate Change Research, University of East Anglia.

4 Slow tourism
Ethics, aesthetics and consumptive values

Michael Clancy

Introduction

Slow tourism has emerged as one segment of an increasingly fragmented tourist market in recent years. Modelled after the slow food and slow city (Cittáslow) movements that first appeared in Italy in the 1980s and 1990s, slow tourism occupies the opposite pole from that of rapid tourist consumption. Dickinson *et al.* (2010:1) define the concept as "an emerging conceptual framework which offers an alternative to air and car travel, where people travel to destinations more slowly overland, stay longer and travel less." Fullagar *et al.* (2012) locate slow travel within the larger slow movement that broadly embraces a more simple and less consumptive lifestyle. The practice of slow travel and tourism[1] is contrasted against the prevailing trend of rapid travel, both as a means of getting from one place to another, and as a form of consumption once there. It contrasts itself, most directly, against the stereotypical package tour where tourists see seven cities in six days before being whisked back home exhausted and overwhelmed by hyperconsumption of sites, or the more recent city break phenomenon, where low cost airlines make it possible to jet abroad for a long, activity filled, weekend.

This chapter examines various aspects of the slow tourism phenomenon. It describes its links to the larger "slow" movement that includes slow food, slow cities and slow living. It then places slow tourism within the larger category of ethical tourism by discussing how the movement promises not only a more fulfilling personal experience for the individual traveler, but also more beneficial consequences associated with this form of tourism. This section also examines ethical tourism within the broader ethical consumption movement. The following section expands the discussion by raising various aesthetic and sociological aspects of consumption itself and the manner in which tourism fits into them. Finally, the concluding section considers the promises and pitfalls of slow tourism in light of these issues.

The emergence of slow tourism

Because of its connection to slow food and slow city movements, most analysts connect slow tourism to the broader opposition to hyperconsumption associated with globalization. Ritzer (1993) locates rapid consumption within his McDonaldsization thesis, where the best practices associated with the fast food chain are applied to many aspects of modern life. Standardization, rationality, convenience and homogenization come to dominate production and consumption patterns over experience, difference and distinctiveness. The slow food movement, which began in 1986 as a protest against the opening of a McDonalds fast food outlet near the Spanish Steps in Rome, pioneered the larger "slow" movement. Although holiday tourism has traditionally been posited as a temporary break from the treadmill of "fast life," it has increasingly become an integral part of that life. Krippendorf (1984), an early critic of mass tourism, anticipated Ritzer's broader critique of modern society. He saw modern Western tourists needing holidays away in order to steel themselves for the next round of hyperactivity through work, consumption and the daily grind. Yet that widespread need for downtime produced what Dickinson and Lumsdon (2010: 6–7) refer to as the traditional tourism system, with its emphasis on maximizing tourist flows, and mass tourism consumption. In such a system the boundaries between work and leisure, fast-paced and slow-paced, begin to break down. Holidays become more and more like non-holidays with tourists forming crowds and moving rapidly as they rush to see sites.

Slow tourism constitutes the antithesis of this system, valuing slowness of pace, enjoying the journey itself, and making connections with local practices and cultures. Travel itself is meant to awaken the senses rather than constitute the precursor to when the "real" holiday begins. And once at the destination, the focus of slow tourism is on tranquility, relaxation and finding meaning by leisurely taking in all that the senses have to offer rather than collecting as many touristic experiences as possible. Slow tourism is connected to an emerging larger critique of speed in all aspects of modern Western life. Dickinson and Lumsdon (2010) argue three guiding principles underlie slow tourism: taking fewer trips, development of low carbon travel, and enrichment of experience. Summarizing Honoré (2004), Lumsdon and McGrath (2011) suggest slow tourism embodies four main principles; experiencing travel itself as part of the holiday; valuing and learning about local culture; slowing down and taking time to relax; and keeping the negative impact on local communities and the larger environment to a minimum. These principles, however, are still emerging. Although slow tourism has been present as a concept and practice for years, organizationally it remains in its infancy, with websites and web communities being formed around the turn of the century and the first international non-profit organization created in 2000. It takes the form of various marketing schemes, travel clubs, and ideas put forth in the popular press. Because the movement has risen more organically, just what constitutes slow tourism remains unclear. Heitmann *et al.* (2011: 117) describe the activity as "a form of tourism that respects local cultures, history and environment, and values social responsibility while celebrating diversity and connecting people (tourists with other tourists and host communities)."

Lumsdon and McGrath (2011) report the results of their survey of several leading thinkers and practitioners of the activity in an effort to gain greater clarity. They found that while many different conceptions of slow tourism exist, there is consensus on three points: slowness, travel experience and concern for the environment. Slowness encompasses pace and the perceived opposition to the cult of speed associated with everyday life. Travel experience is a direct result of this slower pace and claims to produce greater, more in-depth, meaning for the traveler. For some this deeper engagement is the product of travel itself, utilizing alternative forms of transport. Modes such as train travel or cycling in groups are communal and alert the senses. For others, deep engagement is found through a slower pace after arriving at the destination. Dickinson et al. (2011) acknowledge this difference, distinguishing between what they call "hard" slow travelers who are motivated by environmental concerns, and "soft" slow travelers who are guided more by the pleasure of slow. Yurtseven and Kaya (2011) make a similar but slightly different distinction, between what they call Dedicated Slow Tourists, Interested Slow Tourists, and Accidental Slow Tourists. Finally, environmental consciousness is directly related to larger questions of sustainability and concern for the host community. This places slow tourism directly under the rubric of responsible tourism as well as the larger practice of ethical consumption.

Responsible Tourism constitutes a broad category of tourist activities that introduce ethics to tourist practices. It promises both the desired hedonistic outcome for tourists along with desirable social, economic and/or environmental benefits for stakeholders and their communities. A 2002 Cape Town (South Africa) Declaration on Responsible Tourism in Destinations defined responsible tourism as simply "making better places for people to live in, and better places for people to visit" (Responsible Tourism Partnership 2012). Niche tourism labels such as ecotourism, fair-trade tourism, sustainable tourism, community-based tourism, volunteer tourism and pro-poor tourism; all fall under the umbrella of responsible or ethical tourism. At their basis, according to Boluk (2011), lies a concern with turning global tourism into a positive rather than negative force in the world. All these segments reflect a larger moralization of tourism (Butcher 2003; Smith and Duffy 2003; Gibson 2010; Lisle 2009).

Slow tourism shares these concerns in two ways: first is in the mode of travel. Slow tourism encourages "slow travel" from locale to destination. While part of this is aesthetically motivated, slow travel is primarily concerned with impacts, specifically those of greenhouse gas (GHG) emissions and global climate change. As Dickinson and Lumsdon (2010) report, global tourism is estimated to contribute some 4–6 percent of all human-made GHG emissions, and transport usually accounts for 70–90 percent of that total according to Gössling (2002). As a result, Gardner (2009) argues that traveling slow, and avoiding air travel, are crucial to the idea of slow tourism. Although debate continues on whether car travel truly fills the ethical and aesthetic demands of "slow," a concern with the environmental impact of travel remains central to slow tourism.

Second, slow tourism also embodies principles of responsible tourism at the destination. Slow tourists are encouraged to learn about and adopt local cultural customs, to "engage with the community at the right level" (Gardner

2009), "live in harmony with the locality and its inhabitants" (Matos 2004: 100), stay longer, engage more deeply, and be cognizant of their impact on local communities. While consuming local resources is deemed important, it should be done sustainably. As Germann Molz (2010) points out in surveying literature and traveler experience, the very heart of slow tourism contains a desire for a deeper, more meaningful experience while simultaneously doing good in the world.

Slow tourism and the politics of ethical consumption

Alternative consumption, as Sassatelli (2004: 181) summarizes, involves "heterogeneous practices and discourses ... [which] herald a critique of (some forms of) consumption and propose alternative lifestyles." Alternative consumption is social (in that it tends to urge a widespread practice), and ethically motivated. It has grown rapidly in recent years. Initiatives such as fair trade and anti-sweatshop campaigns, boycotts, green consumption, labeling schemes, socially responsible investing and localism, have all gradually been moving to the mainstream. In recent surveys more than 30 percent of residents of several European countries report consumer activism in the form of boycotts, purchasing products based on their social, ethical or environmental qualities, and/or participating in consumer associations (Sassatelli 2006). In the United Kingdom spending on "ethical" products more than tripled between 1999 and 2009 (Barnett *et al.* 2011). These movements have emerged out of consumer awareness surrounding the consequences of mass consumption practices on the climate, working conditions and local cultures, to name a few. The dominant underlying ethical framework is that of utilitarianism, where action is judged morally on its resulting societal balance between pleasure and pain. The promise is that, done correctly, consumers can enjoy their consumer goods while promoting good in the world. These societal movements have resulted in scholars (Micheletti 2003; Micheletti *et al.* 2004; Sassatelli 2004; 2006; De Neve *et al.* 2008; Stehr 2008; Soper *et al.* 2009; Barnett *et al.* 2011) drawing new attention to studies of critical consumption. Among the most pressing questions they have addressed is how do we explain the mainstreaming of these practices in contemporary times, and what do they mean?

As for the former, some scholars place ethical and alternative consumption as a practice directly connected to modernity. The risk society literature suggests late modernity has produced an environmental setting where individuals must negotiate various forms of risk; from what they eat, products they use, where they travel, knowing what toxins they are potentially exposed to. As deregulation and state withdrawal from dealing with these risks increases under neoliberalism, these tasks fall to the individual (Beck 1992, 1997; Almas 1999; Giddens 1999; Micheletti 2003). Distrust of prevailing institutions results in individuals becoming more responsible for finding ways to live "the good life." Part of this is done through daily practices, including consumption. Consumption, traditionally associated with markets rather than politics, now becomes a central arena for the political (Beck 1997; Giddens 1999). This *subpolitics* – because it lies outside of traditional venues of politics – becomes potentially empowering. As Micheletti (2003: 16) puts it: "The focus on choices in daily shopping situations as democratically important implies a new view of consumers as potentially important agents of political change." No longer simply passive responders to corporate marketing, consumers are taken seriously as both economic and political agents. According to this approach we should view "shopping situations as democratically important" (ibid). Adding a social or environmental calculus to shopping and consumption helps transform passive consumers into active citizens, or what Scammell (2000: 16) calls "consumer citizens" in an almost Toquevillian manner. Taken to its furthest conclusion, consumption choice "can be likened to resistance fighters" (ibid), "raises political awareness" and "stimulates democracy" (Worldwatch Institute, cited in Sassatelli, 2006: 221). If true, democratic politics is fundamentally relocated from the ballot box to the marketplace.

This notion of alternative consumption as a new and promising arena for political struggle has, to be sure, come in for criticism. As Sassatelli (2006) points out, not all alternative consumerism is politically motivated, nor is it necessarily stimulated by rationalist motivations. Moreover, as Barnett *et al.* (2011) argue, alternative consumption may constitute a hollow substitution for traditional politics. The larger critique has to do with consumer as sovereign. Maniantes (2001: 33) suggests that moving responsibility for solving collective problems to the individual is highly problematic. This emphasis on ethical consumption as a substitute for traditional action, promotes "the notion that knotty issues of consumption, consumerism, power and responsibility can be resolved neatly and cleanly through enlightened, uncoordinated consumer choice." Ethical consumption further relocates responsibility away from powerful institutions that are often the very cause of social and environmental problems. In fact, the narrative of the consumer as omniscient sovereign is the very product of neoliberalism. Imperfect information, lack of true choice, fragmentation and collective action problems are seldom mentioned. Instead, according to critics, this is a stylized presentation of individual consumer knowledge, rationality and power (Barnett *et al.* 2011).

Missing from this debate is a differentiation between types of ethical consumption. It groups all ethical and responsible consumption together. Is the anti-sweatshop movement the same as Fair Trade coffee or dolphin-safe tuna? There are reasons to believe it is not. Many ethical consumption movements, for instance, either call for abstinence or an ethic of consuming less. The Voluntary Simplicity movement adopts a Spartan aesthetic and urges less overall consumption (Maniantes 2002). This also makes sense for consumption related to environmental issues (global climate change) or scarcity (threatened species). Doing one's part means sacrificing by not eating tuna, or walking rather than driving. Boycotts and much of socially responsible investing also operate through consumer denial. When it comes to tourism, examples include boycotting politically repugnant regimes such as South Africa under Apartheid, or more recently, Myanmar. Tourists may also consciously choose not to travel to environmentally sensitive sites such as the Galapagos Islands. Here, sacrificial denial is the means by which one participates. A second area, substitution, identifies no intrinsic difference between responsible consumer products and "non-responsible" ones. In other words no claims are made regarding ethical bananas or coffee tasting better than ordinary bananas or coffee. The value in this type of consumption, then, is limited to its ethical claim. Fair Trade chocolate is better because it does not use child slaves in its production, and only for that reason. In tourism, certified ecotourism destinations may be more desirable than those simply marketed as ecotourist ones, but not due to intrinsic qualities of the resorts themselves.

This is precisely where the slow consumption movement differs from many other areas of ethical consumption: it weds an intrinsic aesthetic to a consequentialist ethic. Slow is superior to its fast equivalent, *while simultaneously* claiming to be better in terms of social, political and environmental consequences. The Slow Food organization (Slowfood.org) has as its motto, "Good, clean, fair." While originating as a protest against globalization, the movement has come to locate slow food with pleasure, taste and quality. Fast consumption, it argues, is devoid of true pleasure. In contrast, slow food places pleasure back into the equation of everyday experience, "not as a bonus, not as a retreat" (Parkins and Craig 2006: 91) but by reclaiming it as part of daily life. Similarly, proponents hold slow tourism to be superior to "fast" tourism. Gardener (2009) suggests that during the twentieth century speed was increasingly associated with success, but done so without much serious thought. Fast implies an internal cadence that cannot soak in the complexities and nuances of the travel experience. This slower pace of tourism not only benefits the local economy, culture and environment, it is better able to meet the consumptive desires associated with tourism: relaxation, perception, learning, experience, contemplation, wonder and quietude. In sum, the aesthetic claim is that "Slow travel reinvigorates our habits of perception, taunting us to look more deeply into what we thought we already knew" (Gardner 2009: 13).

Aesthetics meet ethical consumption

The discussion thus far surrounds the larger question of consumptive practices. What do we do when we consume? What goes into our desire for certain products? What values are involved? The literature on consumption is vast and beyond the scope of this chapter, but various strands of consumption studies are particularly useful for our understanding of the ethical consumption movement and how and where slow tourism fits in. Immediately, we can distinguish between three types of consumptive values of products: use value, experiential value, and symbolic or social value. Use value refers to direct, intrinsic qualities. We eat because we are hungry and need nutrition. We burn oil because we are cold and heat from the fire provides warmth. Use value satisfies a direct need or desire. Experiential value refers to aesthetics, the broader sensory and perceptual value or consumption. We eat seafood by the coast because the two seemingly go together. We prefer to burn wood over oil because the fire is prettier or smells better. We drive to get from point A to B, but may drive a sports car because of the experience of the wheels hugging the winding road.

Many have noted the growing importance of this experiential and aesthetic aspect of goods and services in affluent societies. Pine and Gilmore (1998) argue that experiences constitute a separate, fourth dimension of consumption (after commodities, goods and services), representing the newest and fastest growing consumer "product" in advanced industrialized societies at the end of the twentieth century. When consumers have their use values largely met, what they desire are experiences in all their forms. Experiences are distinguished from commodities, manufactured goods and services, in that firms "stage" together these products in a new and meaningful way: "Commodities are fungible, goods tangible, services intangible and experiences *memorable*" (1998: 98, emphasis in original). Here the value of products come from what Beckert (2011: 107) refers to as "the fantasies they evoke." Finally, symbolic value refers to the social aspects of consumption. In many cases consumption is related to identity and is meant to differentiate groups in society. We may buy a Prius not for its ability to get from point A to point B but because it identifies us as a member of a group of environmentally concerned consumers.

Together these last two factors – experience and symbolic aspects of consumption – have taken on increased importance in recent years. As Adolph and Steer (2010) point out, a century ago households in advanced industrial societies spent more than 80 percent of their income on the necessities of food, shelter and clothing. Today that figure is between 30–40 percent. Moreover, the amount of choice available to Western consumers, even when it comes to those necessities, leaves a great deal to the more psychological and sociological aspects of consumption. Some have argued that most, perhaps almost all, modern Western consumption today has come to be about experience. Postrel (2003) argues that the look and feel of everyday consumer products hold specific meaning, hence the desire for everything from designer toasters to upscale hardware at Home Depot.

Symbolic values, or the social aspects of consumption, have long been recognized by scholars. Early work by Veblen (1899), and later Bourdieu (1984), understood the social basis for much consumption. For Veblen, conspicuous consumption marked class distinction and both saw taste as innately tied to this distinction. This is directly related to Hirsch's (1976) concept of positional goods, products that derive much of their value from their position of desirability. The value of positional goods derives from exclusivity. The ability of others to enjoy the same products reduces their very desirability. This ties experience and the symbolizing aspects of consumption together. In many cases the aesthetic qualities of consumption, the joys of their experience, are predicated on the exclusion of others.

More recent scholarship on consumption is less class bound and grants greater agency to consumers. For some scholars consumption fulfills a psychological more than a sociological end (Booker 1976; Lears 1983), while others suggest consumer habits are the product of aspirational identity or desire for inclusion in a "consumer micro-culture" (Sirsi *et al.* 1996). These micro-cultures are distinctive, and the product of socially meaningful practices and behaviors. Indeed much of the science of branding is based on this idea of middle class differentiation, where atomistic consumers seek to mark themselves through the consumer products they buy (Haig 2004; Holt 2004; Anholt 2007). As a result, owners of everything from Mini Cooper cars, Harley Davidson motorcycles, Wheaten Terriers to Apple electronics not only enjoy the intrinsic and experiential aspect of those purchases; they also mark themselves as members of a particular "tribe" (Sirsi *et al.* 1996), an imagined community of like-minded consumers.

Tourism is, of course, implicated in this larger consumer dynamic. First and foremost, tourism is an experiential good and a luxury product. Made up of a combination of goods and services, the overall tourism product is the "experience" tourists have while on holiday. This experience is opposed to that of the everyday, including work and home life. Its hedonistic aesthetic, therefore, is the novel, the unique. Krippendorf (1984: xv), an early critic of mass tourism, referred to it as part of the modern "recreational cycle" where "we travel in order to recharge the batteries, to restore our physical and mental strength .... We then return home, more or less fit to defy everyday life until next time."

Tourism destinations have long been marketed as the converse of home. Middle class travelers are promised luxury and an escape from the everyday. Yet returns diminish as more and more people desire this rejuvenation through tourism. Traveling among the hoards is no longer novel or relaxing. This point is of central importance because it acknowledges the aesthetics of consumption take place in a social setting. Put differently, tourism is often a positional good because the ability to enjoy its fruits is predicated on the exclusion of others. In the modern tourism system, tourists seek to distinguish themselves through not only their choice of destination, but increasingly through the particular choice of niche tourism product. This helps to explain the vast segmentation of tourism markets over the past two decades. Adventure tourists, ecotourists and volunteer tourists, to name a few, mark their identity in society through these consumptive choices.

Slow tourism: wedding the aesthetic to the ethical

Ethical consumption in general, and ethical tourism in particular, can and should be seen within this larger discussion of consumptive practices. The use value of ethical consumption is intrinsic to the product, but the experiential aspect now includes the feeling of virtue (and membership among the virtuous). As an official at UK ethical tourism NGO "Tourism Concern" summarizes: "You can have a great holiday and ... not take a guilt trip!" (quoted in Gibson 2010: 523). Soper's discussion of "alternative hedonism" is of some use here. She distinguishes it from purely ethical consumption due to the primacy of its intrinsic or aesthetic value. "Alternative hedonism represents a critical approach to contemporary consumer culture that is distinctive in its concern with self-interested rather than altruistic motives for shifting to greener lifestyles" (Soper 2009: 4). Originating as much from disaffection with the promise of Western consumer affluence than with questions of social justice, the movement is ultimately motivated by alternative desires associated with "the good life" (ibid: 5). The alternative hedonism model takes a middle ground in its conception of the consumer, envisioning her as reflexive but existing within a larger environment "immersed in consumer culture" (ibid: 11). Slow living fits directly within this model of alternative consumption in that it is motivated not just by external concerns but also this desire for right living, regaining agency over everyday life (Parkins and Craig 2006: 67), and seeking deep pleasure: the very heart of alternative hedonism.

The discussion above suggests that while the experiential and symbolic aspects of consumption may be distinctive, they are not wholly separate from one another. This holds importance for ethical tourism in general, and slow tourism in particular. Much of the appeal of consuming ethically is not simply the consequences of that consumption on others or the environment; it is also in the aesthetic value of feeling virtuous. Saving the environment, helping the poor, or achieving some kind of distributional justice, become qualities associated with consumer products. Ethical tourism holds precisely this promise; tourists cannot only have their holiday, with all the aesthetic values holidays include, but also enjoy the additional experience, however temporary, of feeling principled and moralistic. This promise that all good things can go together is often accepted uncritically. As Lisle (2010: 142) puts it rather sarcastically, "not only does ethical tourism solve all the problems caused by mass tourism, it also makes you a better person and the world a better place." She and others (Butcher 2003; Hall 2007; Harrison 2008; Guttentag 2009; Gibson 2010; Vrasti 2010) criticize dominant mainstream models of ethical tourism, arguing they simply add the issue of ethics to existing tourism uncritically.

This raises the issue of whether and how these different values fit together – or do not. Vrasti's (2010) study of volunteer tourism points to the difficulty of tying hedonistic pleasure to virtue. Her interviews and observations of "voluntourists" to Guatemala and Ghana show that many became quickly disillusioned by the volunteer aspect of their trips and in many cases chose activities associated with hedonism over those associated with helping local communities. Here the slow tourism movement is similar to other areas of ethical tourism but also contains important differences. Predominant forms of ethical tourism such as sustainable tourism, pro-poor tourism, ecotourism and volunteer tourism promise social, economic and/or environmental benefits, usually promised as *additions to* the usual desired components of a tourist product such as seeing a new place, learning about a different culture, and self-reflection. Aside from the added aesthetic of doing good, however, these hedonistic desires are not necessarily qualitatively different than mainstream tourism. In many cases, in fact, there is a direct trade-off. Voluntourism, for instance, weds travel and leisure with performing labor, hence invoking sacrifice. Slow tourism, in contrast, not only doesn't invoke denial or sacrifice; it makes additional *aesthetic* claims; it promises to be intrinsically better than fast tourism. The slow experience is superior to other forms of touristic experience, and has the added ethical benefit of avoiding the negative social, economic and environmental effects often associated with mainstream tourism.

Yet there is some reason to ask whether these two will always exist in harmony. In a recent article, Slow Food USA President, Josh Viertel (2012), raises precisely this issue. While originally formed as a protest social justice movement against the encroachment of fast food, much of the subsequent growth of the slow food movement owes to the rediscovery of craft-made, locally produced quality food, rediscovered cuisines and ingredients, and above all the simple pleasures of consuming good food and drink. A crucial element of the movement has been an effort to link consumer and producer in solidarity. Viertel, a former farmer who along with his wife sold his products in local farmers markets to foodies at high prices, points out he made just $12,000 a year doing so. The slow food movement actually prices out the poor and fails to provide a living wage for producers, and Viertel suggests that much of the movement has come to focus on the pleasures of eating at the expense of addressing more difficult social questions: "Can you both fight for the farmer and fight for the eater, or do farmers and eaters have competing agendas?" (Viertel 2012).

Similar questions are likely to emerge with respect to slow tourism as the activity grows. Although in its infancy, thus far much of the literature on the activity claims an identity between aesthetics and ethics. Do they *really* coincide? Is that balance *always* likely to be present? Is slow tourism accessible to broad segments of the population, or only to the most affluent? What are the carrying capacity limits to a slow destination, not simply from a sustainability standpoint, but also from an aesthetic one? Can slow tourism flourish on a large scale? Does it need to? Spatially can it co-exist with mass tourism? Other practical problems also exist, and they can be best seen when compared to slow food and slow cities. Both of those movements have potential communities in place. Most slow food movements are community oriented and focus on the local. Slow cities are based on a community relationship as well. Tourists, however, are transient and their contribution to establishing community ties is time limited. They are also by nature outsiders. Will locals trust that they have shared interests?

It is important to note that most of the authorship of the slow tourism movement has come from the tourist, or at least is written in their name. At issue is how travelers should travel, what they should consume, and the pace at which they do so. What is assumed in all of this is what is good for the tourist is good for the tourism provider and the broader host community. Little attention, it seems, has been paid to these groups. If alternative consumption truly is political, the polity must be inclusive. Instead, as with much of ethical tourism, solidarity is assumed between slow tourist, tourism provider and host community. But this solidarity is potentially one way, and is dependent on the desirability of a particular host market. This is also true with respect to which destinations are deemed as sufficiently slow to attract slow tourism. If slow tourism shares patterns of other tourism markets, the fragility of destinations due to tourism being a positional good remains a concern. Part of the sustainability ethic refers not just to protecting resources but also maintaining livelihoods over the long term. Some (Matos 2004; Conway 2010) have argued that various destinations can rebrand themselves as slow markets, giving some voice to stakeholders, but how far this can go – both in terms of giving a voice to the host community and whether slow tourism can move from being a niche tourism product to a true political and social movement – remains unclear.

Further information

www.slowtrav.com/
www.slow-tourism.net/
www.slowtourismclub.eu/main/
www.slowmovement.com/

Activity

Among the central questions associated with slow tourism and ethical consumption are definitions of the phenomenon, intrinsic attractiveness of the activity, and benefits associated with it.

For the activity, divide the class into two groups: potential slow tourists and potential slow tourism providers. The groups are to discuss the following questions:

- What is slow tourism? What are the main components of the activity?
- Why would your group engage in this activity as opposed to another form of tourism?
- What are the primary benefits you are seeking from the activity?
- What makes this activity "ethical?"

After considering these questions the two groups should reconvene and compare their answers. The goal is to highlight the commonalities and differences among the groups and to focus on their respective interests. Further discussion can center upon whether and how the differences can be reconciled.

Note

1 Some distinguish between slow travel and slow tourism, but I use them interchangeably here.

References

Alter, L. (2008) "The Slow Movement Isn't Just About Food" in *Huffington Post*, August 31. Online, available at: www.huffingtonpost.com/lloyd-alter/the-slow-movement-isntju_b_122792.html (accessed January 7, 2012).

Almus, R. (1999) "Food Trust, Ethics And Safety In Risk Society" in *Sociological Research Online*, 4 (3). Online, available at: www.socresonline.org.uk/4/3/almas.html (accessed January 7, 2012).
Anholt. S. (2003) *Brand New Justice: The Upside Of Global Branding*, London: Butterworth-Heinemann.
Barnett, C., Cloke, P., Clarke, N. and Malpass, A. (2011) *Globalising Responsibility: The Political Rationalities Of Ethical Consumption*, Chichester: Wiley-Blackwell.
Beck. U. (1992) *Risk Society: Toward A New Modernity*, London: Sage.
Beck. U. (1997) *The Reinvention Of Politics: Rethinking Modernity In The Global Social Order*, Oxford: Polity Press.
Beckert, J. (2011) "The Transcending Power Of Goods: Imaginitive Value In The Economy" in J. Berkert and P. Aspers (eds.), *The Worth Of Goods: Valuation And Pricing In The Economy*, Oxford: Oxford University Press, 106–28.
Boluk, K. (2011) "In Consideration Of A *New* Approach To Tourism: A Critical Review Of Fair Trade Tourism" in *The Journal of Tourism and Peace Research*, 2 (1): 27–37.
Bourdieu, P. (1984) *Distinction: A Social Critique Of The Judgment Of Taste*, Cambridge, MA: Harvard University Press.
Butcher, J. (2003) *The Moralisation Of Tourism: Sun, Sand ... And Saving The World?* London: Routledge.
Castells, M. (1996) *The Rise of the Network Society*, Cambridge, MA: Blackwell.
Cittáslow (2012) Cittaslow official list. Online, available at: www.cittaslow.org/download/DocumentiUfficiali/CITTASLOW_LIST_12:2011.pdf (accessed January 4, 2012).
De Neve, G., Luetchford, P., Pratt, J. and Wood, D.C. (eds.), (2008) *Hidden Hands of the Market: Ethnographies Of Fair Trade, Ethical Consumption, And Corporate Social Responsibility*, Research In Economic Anthropology 28. Bingley, UK: JAI Press.
Dickinson, J.E. and Lumsdon, L. (2010) *Slow Travel and Tourism*, London: Earthscan.
Dickinson, J.E., Lumsdon, L. and Robbins, D.K. (2011) "Slow Travel: Issues For Tourism And Climate Change" in *Journal of Sustainable Tourism*, 19 (3): 281–300.
Dickinson, J.E., Robbins, D. and Lumsdon, L. (2010) "Holiday Travel Discourses And Climate Change" in *Journal of Transport Geography*, 18: 482–9.
Fullagar, S., Wilson, E. and Markwell, K. (2012) "Starting Slow: Thinking Through Slow Mobilities And Experiences" in S. Fullagar, K. Markwell and S. Wilson (eds.), *Slow Tourism: Experiences And Mobilities*, Clevedon: Channel View Publications: 1–8.
Gardner, N. (2009) "A Manifesto For Slow Travel" in *Hidden Europe Magazine*, 25: 10–14.
Germann Molz, J. (2009) "Representing Pace In Tourism Mobilities: Staycations, Slow Travel And 'The Amazing Race'" in *Journal of Tourism and Cultural Change*, 7 (4): 270–86.
Gibson, C. (2010) "Geographies Of Tourism: (Un)-Ethical Encounters" in *Progress in Human Geography*, 34(4): 521–7.
Giddens, A. (1999) *Runaway World: How Globalization Is Reshaping Our Lives*, London: Routledge.
Gleick, J. (1999) *Faster: The Acceleration Of Just About Everything*, New York: Pantheon.
Goodwin, H. (2011) *Taking Responsibility for Tourism*, Oxford: Goodfellow Publishers.
Gössling, S. (2002) "Global Environmental Consequences For Tourism" in *Global Environmental Change*, 12: 283–302.
Guttentag, D. (2009) "The Possible Negative Impacts Of Volunteer Tourism" in *International Journal of Tourism Research*, 11 (6): 537–51.

Haig, M. (2004) *Brand Royalty: How The World's Top Brands Thrive And Survive*, London: Kogan Page.
Hall, C.M. (ed.) (2007) *Pro-Poor Tourism: Who Benefits?* Clevedon: Channel View Publications.
Harrison, D. (2008) "Pro-Poor Tourism: A Critique" in *Third World Quarterly*, 29 (5): 851–68.
Heitmann, S., Robinson, P. and Povey, G. (2011) "Slow Food, Slow Cities And Slow Tourism" in P. Robinson, S. Heitmann and P.J.C. Dickey (eds.), *Research Themes for Tourism*, London: CAB International, 114–27.
Hirsch, F. (1977) *The Social Limits to Growth*, London: Routledge.
Holt, D.B. (2004) *How Brands Become Icons: The Principles Of Cultural Branding*, Boston: Harvard Business School Press.
Honoré, C. (2004) *In Praise of Slowness: How A Worldwide Movement Is Changing The Cult Of Speed*, San Francisco: Harper Collins.
Krippendorf, J. (1984) *The Holiday Makers*, London: Heinemann.
Lears, T.J.J. (1983) "From Salvation To Self-Realization: Advertising And The Therapeutic Roots Of The Consumer Culture, 1880–1930" in R.W. Fox and T. J. J. Lears (eds.), *The Culture of Consumption: Critical Essays in American History, 1880–1980*, New York: Pantheon, 1–38.
Lisle, D. (2009) "Joyless Cosmopolitans: The Moral Economy Of Ethical Tourism" in M. Paterson and J. Best (eds.), *Cultural Political Economy*, London: Routledge, 139–57.
Lumsdon, L.M, and McGrath, P. (2011) "Developing A Conceptual Framework For Slow Travel: A Grounded Theory Approach" in *Journal of Sustainable Tourism*, 19 (3): 265–79.
Maniantes, M. (2001) "Individualization: Plant A Tree, Buy A Bike, Save The World?" in *Global Environmental Politics*, 1 (3): 31–52.
Maniantes, M. (2002) "In Search Of Consumptive Resistance: The Voluntary Simplicity Movement" in T. Princen, M. Maniantes and K. Conca (eds.), *Confronting Consumption*, Cambridge: MIT Press: 199–235.
Matos, R. (2004) "Can Slow Tourism Bring New Life To Alpine Regions?" in K. Weiermair and C. Mathies (eds.), *The Tourism and Leisure Industry: Shaping The Future*, Binghamton, NY: Haworth Hospitality Press: 93–103.
Micheletti, M. (2003) *Political Virtue and Shopping: Individuals, Consumerism, And Collective Action*, London: Palgrave MacMillan.
Micheletti, M., Follesdal, A. and Stolle, D. (eds.) (2004) *Politics, Products And Markets: Exploring Political Consumerism Past And Present*, New Brunswick, NJ: Transaction Publishers.
Parkins W. and Craig, G. (2006) *Slow Living*, Oxford: Berg.
Petrini, C. (2001) *Slow Food: The Case For Taste*, New York: Columbia University Press.
Pine II, B.J. and Gilmore, J.H., (1998) "Welcome To The Experience Economy" in *Harvard Business Review*, 76 (4): 97–105.
Postrel, V. (2003) *The Substance Of Style: How The Rise Of Aesthetic Value Is Remaking Commerce, Culture And Consciousness*, New York: Harper Collins.
Responsible Tourism Partnership (2012) "The Cape Town Declaration." Online, available at: www.responsibletourismpartnership.org/CapeTown.html (accessed January 5, 2012).
Ritzer, G. (1993) *The McDonaldization of Society*, Thousand Oaks, CA: Pine Forge Press.

Sassatelli, R. (2004) "The Political Morality Of Food: Discourses, Contestation And Alternative Consumption" in M. Harvey, A. McMeeckin and A. Warde (eds.), *Qualities of Food: Alternative Theoretical And Empirical Approaches*, Manchester: Manchester University Press: 176–91.

Sassatelli, R. (2006) "Virtue, Responsibility And Consumer Choice: Framing Critical Consumerism" in J. Brewer and F. Trentmann (eds.), *Consuming Cultures, Global Perspectives: Historical Trajectories, Transnational Exchanges*, Oxford: Berg: 219–50.

Scammell, M. (2000) "The Internet And Civic Engagement: The Age Of The Political Consumer" in *Political Communication*, 17: 351–5.

Sirsi, A.K., Ward, J.C. and Reingen, P.H. (1996) "Microcultural Analysis Of Variation In Sharing Of Causal Reasoning About Behaviour" in *Journal of Consumer Research*, 22: 345–72.

Smith, M. and Duffy, R. (2003) *The Ethics of Tourism Development*, London: Routledge.

Soper, K. (2009) "Introduction: The Mainstreaming Of Counter-Consumerist Concern" in K. Soper, M. Ryle and L. Thomas (eds.), *The Politics and Pleasures of Consuming Differently*, Basingstoke: Palgrave MacMillan: 1–21.

Tomlinson, J. (1999) *Globalization and Culture*, London: Polity.

Worldwatch Institute (2004) *State Of The World 2004: A Worldwatch Institute Report On Progress Toward Sustainable Society*, Washington: Worldwatch.

Vrasti, W. (2010) "The Self As Enterprise: Volunteer Tourism In The Global South," Unpublished Thesis, McMaster University.

Viertel, J. (2012) "The Soul Of Slow Food: Fighting For Both Farmers And Eaters" in The Atlantic Online, online, available at: www.theatlantic.com/health/archive/2012/01/the-soul-of-slow-food-fighting-for-both-farmers-and-eaters/251739/ (accessed January 24, 2012).

Yurtseven, H.R. and Kaya, O. (2011) "Slow Tourists: A Comparative Research Based On Cittáslow Principles" in *American International Journal of Contemporary Research*, 1 (2): 91–8.

5 The evolution of environmental ethics
Reflections on tourism consumption

Andrew Holden

Introduction

The focus of this chapter is to consider the relationship of the consumption of tourism to environmental ethics in the context of the debate on the limits to consumerism. Human desires are seemingly insatiable, posing a serious challenge to the ability of environmental systems to be able to accommodate them and over two decades after the 'Earth Summit' in Rio de Janeiro, what may have then seemed far-off concerns have now become reality; including climate change, an accelerating loss of biodiversity, desertification and land degradation. The growth in tourism consumerism is illustrated by the increase in recorded international tourist arrivals from 25 million in 1950 to 947 million in 2011 (UNWTO 2012). It is this growth in recreational mobility and the extension of tourism's spatial peripheries that has led to an increased recognition of its use for economic development. More recently there have been calls for tourism to be a key economic sector in a future green economy, one that aims to create 'improved human well-being and social equity, while significantly reducing environmental risks and ecological scarcities' (UNEP 2011: 2).

As UNEP (ibid.) recognize, critical to meeting this aim is that tourism is 'well designed' to interact beneficially with the local economy, reduce poverty and enhance nature conservation. It is evident that in situations where tourism can lend an economic value to the biodiversity of nature through visitation, there exists a more persuasive argument for environmental conservation, as is discussed in the Mabira Rain Forest case study at the end of this chapter. However, as is exemplified at Mabira, tourism may face competition from claims of other 'environmentally friendly' industries for land use: in this case biofuels. Nor is the challenge of a well-designed tourism a straightforward one given the complexity of a system that links distant spatial areas and involves multiple stakeholders. The key driver of the system is consumer demand, and subsequently it is argued that the environmental behaviour of tourists will be a significant determinant of tourism's relationship with nature. This challenge is emphasized in the UNEP's (2011) identification of five key challenges to tourism's place in the green economy, four of which relate to the natural environment and, to differing degrees, to aspects of consumer behaviour. These are a consumer trend towards travelling longer distances for shorter durations of time; a preference for energy intensive transport based on carbon fuels, and an associated growth in GHG emissions; excessive water consumption; and damage to marine and terrestrial biodiversity.

The challenge of how to solve environmentally damaging consumer behaviour is one that can be approached from varied perspectives embracing the economic, philosophical and technological. Environmental economists emphasize the necessity for correction of market failures that have led to the overuse and depletion of scarce resources and pollution, typically through price adjustments that reflect the full environmental and social costs of their use, theoretically providing a powerful incentive to change behaviour. Faith may also be placed on continued advancements in green technologies to produce solutions to environmental problems that will necessitate a minimum interference to our lifestyles and avoid a re-evaluation of lifestyles, often the most politically acceptable solution to dealing with environmental problems. While knowledge advancements in technology and economics have an important role to play in a green future, an important caveat, as Johnston (2008) points out, is that solutions are being sought from systems that have encouraged anthropocentric-driven environmental problems.

Without alterations to consumer behaviour, technological solutions may prove inadequate when faced with an increased demand for services or goods that are environmentally damaging. This point is exemplified through the seeking of a technological solution to solve the challenge of mitigating the increasing contribution of aviation to total GHG emissions. Driven by increasing demand in developed countries for low cost, long haul flights over the last two decades, and economic globalization that continues to rapidly expand the middle classes of the BRIC (Brazil, Russia, India and China) and other developing countries, this has led to a global propensity for air travel and the numbers of air passengers continues to grow exponentially. It is this growth in demand that reduces the opportunity for technological innovation to reduce the total levels of global GHG emissions from aviation, despite marked advancements that continue to reduce emissions from individual aircraft engines.

An alternative technological solution to the challenge of aviation derived GHG emissions lies in a switch from a reliance on carbon based fuels to non-environmentally polluting alternatives. While this may initially seem an attractive option, there are economic and environmental consequences. For example, the growing of biofuel crops for aviation, when technologically feasible, may have negative effects on food supply as agricultural land is switched to alternative use, causing food prices to rise and subsequently disproportionately disadvantaging the world's poor. The demand for land for biofuel agriculture may also cause a reduction in biodiversity, as is explained in this chapter's case study.

It is subsequently argued that it is impossible to decouple the dominant environmental worldview of a society from economic and technological innovation as a determinant factor in how it interacts with and impacts upon nature.

This view is inherent to environmental philosophy, an offshoot of mainstream philosophy that has had increased attention since the 1970s as the ability to deal with environmental problems within current environmental, economic and philosophical constructs has been increasingly challenged.

Environmental philosophy and ethics

Discussion of the relevance of environmental philosophy and ethics to the consumption of tourism is indicative of a wider re-evaluation of our place in nature and our relationship to the environment. It is also symptomatic of a recognition that many of the environmental problems identified by science, including climate change, ozone depletion and pollution, have anthropocentric (human derived) origins. The science of ecology has lent a greater understanding of the interrelationship and co-dependency of natural species, and a subsequent emphasis upon us being a part of nature rather than separate from it. This incorporates a comprehension that human created economic and social processes that are decoupled from nature are unsustainable.

The re-evaluation of our position relative to the natural environment extends to consideration of our duties and obligations to the non-human environment and the judgement of the rights or wrongs of our actions towards it. While it may initially seem strange to think of ethical relationships extending beyond the boundaries of human concerns, this can be understood within the context of the evolution of the concept of community and associated sentiments of empathy and sympathy. While Charles Darwin recognized the importance of the sentiments of empathy and sympathy for human co-operation and survival, Wenz (2001) emphasizes that they are not extended equally to all individuals, typically being stronger among family and friends where dependency and familiarity is at its highest. However, economic and social forces continue to create increasing dependency and familiarity between places – encapsulated in the concenpt of the 'global village'. It can be argued that the resultant heightened inter-connectivity brought about by globalization has led to an increased mutual awareness and sentimental concern between peoples, as evidenced in global responses to natural disasters such as the Asian Tsunami, successive famines in several African countries, and also calls from the United Nations for universal human rights.

Reflecting on the spatial limits of community being extended to people we may never actually physically interact with but with whom we have an empathetic connection, it is possible to envisage the concept of community extending to the non-human world as we re-evaluate our position in nature. Central to defining ethical relationships within a community is the issue of rights, which can be envisaged as an evolutionary continuum. Examples include the abolition of feudalism and slavery, the United Nations Declaration of Human Rights, gender equality, legislation to arrest the persecution of minority groups based on their race, ethnicity, and sexuality; and, significantly in the context of the non-human environment, the extension of legal protection to animals. These are all examples of an evolutionary process extending ethical reasoning to a wider community.

While the concept of an environmental ethic can be understood in the context of present environmental concerns, the legacy of contemporary environmentalism can be traced to the impact of the Industrial Revolution on seminal environmental thinkers, including Henry Thoreau and John Muir. Concerned that industrialization and urbanization in nineteenth century USA would lead to loss of the wilderness he loved so much, Thoreau was highly critical of society's values as compared to those embedded in the spirit of nature. This criticism extended to the capitalist system itself, which he believed would result in the privatization of wilderness, restricting access to nature for ordinary people. An environmental ethic, with humans as a part of nature rather than separate from it, is put forward by Thoreau (1993: 49):

> I wish to speak a word for Nature, for absolute freedom and wilderness, as contrasted with a freedom and culture merely civil, – to regard man as an inhabitant, or a part and parcel of Nature, rather than a member of society.

Inspired by Thoreau's philosophy, John Muir was instrumental in the growth of contemporary environmentalism, notably through his activities in founding the Sierra Club in 1898 and the protected area movement in the USA. Muir held wilderness and nature as the source of humanity's spiritual health. He believed that through immersion in nature, and spending time on the land, one could learn how to live best and develop wisdom. Muir was instrumental in the establishment of national parks as a way of enabling urban dwellers to reconnect with nature and its values, and as a way of imparting an ecological education.

Building on the ideas of Thoreau in establishing a relationship between humans and their natural environment, set within an ethical framework, is the writing of Aldo Leopold (1949). In his concept of the 'land ethic' he calls for an acknowledgement of the role of humans as members of the ecological community rather than as conquerors of it. In contrast to Muir, who encouraged visiting the wilderness, Leopold (1949) was openly critical of the travel trade, which he viewed as promoting the sort of mass access to nature that would consequently reduce opportunity for solitude and a 'true' wilderness experience. Writing half a century later than Muir, Leopold noted (1949: 165) that the development of the railways, followed by increased car ownership, had made the experience of solitude in the wilderness more difficult to attain:

> Recreation became a problem with a name in the days of the elder Roosevelt, when the railroads which had banished the countryside from the city regions began to carry city-dwellers *en masse*, to the countryside. It began to be noticed that the greater the exodus, the smaller the per-capita ration of peace, solitude, wildlife and scenery, and the longer the migration to reach them.

Although Leopold's work was perhaps not as popular at the time it was written as it is now (in his own time the major global preoccupation was with post-Second World War reconstruction), his works have since gained increasing recognition both within and outside the environmental movement.

While the works of Thoreau, Muir and Leopold (either implicitly or explicitly) recognize a need for an environmental ethic to guide human interactions with nature, the academic debate about environmental ethics as part of an environmental philosophy has emerged strongly during the last four decades, encompassing a diversity of worldviews and ethical reasoning concerning the rights of nature. The mere discussion of an environmental ethic represents a significant watershed from the perspective of 'Instrumentalism', a philosophy that views nature as having no right to existence. Subsequently it is argued that human impacts upon nature are unworthy of consideration, aside from their effects on other humans, a stance supported by dualist Cartesian philosophy that stresses our moral superiority to other beings (Nash 1989). The position of the non-recognition of the rights of nature is rooted in normative anthropocentrism, a belief that a moral standing is limited to human beings and holds that nothing except human well-being has any intrinsic value. This inherent principle of human superiority over nature offers a rationale for the use of nature for our own benefit, without consideration for any subsequent impact on biodiversity or upon other species. Transferring this system of reasoning to the context of tourism, we are thus free to use nature how we want, aside from consideration of how our actions may affect other people. In the present era of heightened environmental awareness this viewpoint is increasingly less acceptable, but is far from being eradicated. It can be argued, for example, that the Plan Nacional de Estabilización for tourism development in Spain in 1959 – which emphasized 'crecimiento al cualquier precio' or 'growth at any price' – is an example of harnessing nature for our own ends.

A belief that nature has no intrinsic value means that our relationship with it is determined only by how our actions affect the interests and rights of other people. Thus it can understood within the context of the three major ethical traditions related to the 'rights' and 'wrongs' of human behaviour, i.e. 'Egoism', 'Utilitarianism' and 'Altruism'. While these three traditions have different stances on the 'rights' and 'wrongs' of particular actions, all are concerned solely with the morality of human interactions and obligations, ignoring all ethical concerns or claims that relate to the non-human world. Egoism is based upon the view that we will behave in ways which will be beneficial to the self; when faced with moral choices the egoist will select the option he thinks most likely to produce the most benefits for himself. Transposing this to an interpretation of how the natural environment should be treated, emphasis would be placed on the maximization of self-interest. This does not necessarily mean that nature would automatically be targeted for destruction; an individual may very well decide that they will maximize their own benefit by conserving resources rather than destroying them. In the case of a safari-type tourism based on hunting, the egoist would be likely to maximize the pleasure of killing as many animals as possible, unless they perceived that the animal population was being reduced to such a level that it posed a serious threat to hunting in the future.

Rather than focussing on what is best for one's self, utilitarianism, the second ethical tradition, makes judgements of what is 'right' and 'wrong' based upon the extent to which an action benefits the *majority* rather than just one person. The aim of utilitarianism is subsequently to maximize the happiness or pleasure for the greatest number. Extending this ethical principle to human relationships with the environment, it emphasizes that the extent to which the use or non-use of nature can be judged to be right is determined by the extent to which it brings benefits for the majority of people. So, for example, if it was believed that the killing of animals as part of safari tourism was upsetting to the majority of people, utilitarianism would support a ban; not on the principle that it is detrimental to the welfare of animals but rather that it harms the well-being of the majority of other humans. At the opposite end of the moral scale to egoism is altruism, which holds that when making ethical decisions we should put the needs and wants of other people before our own; in this sense it is 'selfless'. As noble as this may seem, it does not extend to consideration of the well-being of the non-human environment, thus altruism exists only within the boundaries of the human world and not beyond it. In the safari tourism and hunting case, the altruist would be concerned with prioritizing the enjoyment of fellow participants above their own rather than being concerned in any way about the the well-being of the animals.

The antithesis of instrumentalism is the paradigm of 'deep ecology' formulated by Naess (1973). This embodies an ethical reasoning in which all sentient and non-sentient beings have a right 'to be'; they have an intrinsic value in their own right. This right 'to be' bestows upon them a right to existence, independent of any value given by humankind. The recognition of an intrinsic value of nature subsequently raises issues about the 'rights' of nature, the basis upon which these rights are decided, and the extent and depth of them. These issues were central to the work of Stone (1972) who advocated for the rights and legal status of sentient and non-sentient beings in the context of a legal debate about representation of nature:

> It is not inevitable, nor is it wise, that natural objects should have no rights to seek redress on their own behalf. It is no answer to say that streams and forests cannot have standing because streams and forests cannot speak.

Certainly advancements in legal legislation such as a 'Wild law' or 'Earth jurisprudence', concepts that recognize the rights of an 'Earth community' in which humans as part of that community cannot ignore the rights of the rest of it (Thornton 2007), would add a new dimension to how we approach developmental decision-making. For example, under this law, a hotel owner could be sued for damages by those acting on behalf of species whose ecological habitat was being destroyed by effluent discharged into the sea by the hotel in question. If the case was won, other means would have to be found for disposal of the sewage; the right to existence of the species of plants and animals affected would be paramount.

Adoption of the principles of deep ecology in policy and legislation could pose a major challenge to our patterns of behaviour, raising issues of our right to use nature in any way for our own benefit, including for food. There do, however, exist in both religious practice and within secular society, various continuums and degrees concerning the recognition of the intrinsic rights of nature – Jainism, vegetarianism and veganism practice the non-consumption of certain resources and minimal consumption of others. Similarly, Naess (1989) accepted that human participation in nature will inevitably involve some element of killing, which he argued as being ethically acceptable as long as it was conducted to satisfy essential needs.

The challenges of defining definitive ethical principles to establish the rights of nature means that the judgement of what is 'right' or 'wrong' in the 'use' of the environment invariably falls into the classification of 'situation ethics', with each situation being judged upon its own merits. The most prevalent overarching contemporary ethic guiding human interaction with nature is the conservation ethic (Vardy and Grosch 1999), which can be interpreted as operating at both anthropocentric and non-anthropocentric levels, i.e. the conservation of nature for human benefit, as well as conservation based on the recognition of the rights of nature to an existence. The conservation ethic is central to the key policies of sustainable development and a green economy, but also within the context of tourism. Holden (2003) suggests that it represents the moral reasoning of most tourism stakeholders. Emphasis is often placed on conservation for *human* benefit as demonstrated in the United Nations (UNWTO 2013) 'Global Code of Ethics for Tourism' (GCET), Article 3, which states: 'All the stakeholders in tourism development should safeguard the natural environment with a view to achieving sound, continuous and sustainable economic growth satisfying equitably the needs and aspirations of future generations.' The conservation ethic is also inherent in initiatives towards Corporate Social Responsibility (CSR) by tourism multinationals, the benefits of which are summarized by UNEP (2005: 8) as offering 'significant business advantages for a company, in terms of its cost savings, market share, reputation and preservation of its main business assets – the places and cultures their clients are willing to pay to visit'. Reference to the intrinsic value of nature is absent in this statement; instead nature is treated as an asset, and conservation as a force for profit maximization.

The extent that a conservation ethic influences tourism consumer behaviour is contestable. While international agencies such as UNEP (2011) emphasize increasing consumer ethical awareness as a key driver for the greening of the tourism industry, the empirical evidence is contradictory. For example, while UNEP (2011) cites a worldwide survey of travellers conducted by TripAdvisor that found 38 per cent of respondents had stayed at an environmentally friendly hotel and that environmentally friendly tourism was a consideration in the choice of holiday, a comprehensive environmental attitudinal survey of over 1,000 households in the UK found that although 80 per cent of households accepted that climate change would affect them, and the same number believed that climate change was already having an effect, only 22 per cent were willing to fly less (Energy Saving Trust 2007). When queried about their willingness to make personal sacrifices for the sake of the environment, flying less was the second most unpopular choice out of five alternatives presented. Similarly, an unwillingness to make a behavioural change to fly less was revealed in Becken's (2007) research in New Zealand. She found a demonstrable reluctance by tourists to take voluntary initiatives and be proactive in addressing the global impact of air travel. Likewise, Hares *et al.* (2010), in their research into holiday decision-making factors in the UK, found that an awareness of climate change did not influence the consumption of aviation, despite the fact that flying was identified as the third most common factor in how an individual's lifestyle impacted upon climate change.

Just as the impact of flying on climate change can be understood as an incremental and cumulative process, the behaviour of tourists in destination environments is influential for determining the outcomes of the tourism–nature relationship. Even in types of tourism that would appear to be ecologically focused, a shared environmental ethic may not be held by all participants as exemplified in the following statement based upon a nature-based experience to Kilimanjaro:

> I went on a trek up Kilimanjaro in Tanzania, and what struck me was the amount of rubbish there. We were expressly told not to leave any rubbish, and yet people did. I was disgusted that people didn't listen. You feel that you put the effort in – why can't everyone else do the same?
>
> (Chesshyre 2005: 4)

This lack of a shared environmental ethic is reinforced in empirical research conducted by Zografos and Allcroft (2007) into the environmental values and attitudes of people who are regular visitors to areas of natural beauty. They discovered that there were high degrees of variance between visitors, relating to: attitudes and behaviour towards nature; levels of support for species equality; degrees of concern for the well-being of the natural resources and ecosystems of the earth and associated environmental limits; and belief in the level of human skills and development to deal with environmental problems. An illustration of how an environmental ethic can influence tourist behaviour at a local level can be imagined in the following example. The first visitor to a forest, recognizing the intrinsic value of nature and wildlife around them, avoids making an unnecessary disturbance out of respect for other sentient beings, even when it is certain there are no other people around. In contrast, a second visitor who places little value on 'wilderness' (beyond its anthropocentric utility) may also choose to not make unnecessary noise or create disturbance *if other humans are present* but might otherwise feel at liberty to make as much noise as they wish.

Conclusion

While the idea of the relationship of an environmental ethic to tourism may seem abstract, environmental awareness has begun to influence consumption both at a conscious and subconscious level. As we have become aware of the environmental challenges we face, and search for solutions to mitigate and adapt to them, a re-evaluation of our place in nature and our relationship to the surrounding environment is inevitable. Although adjustments to economic systems to incorporate environmental costs and technological developments are an essential part of this process, it is uncertain they can provide full and sustainable solutions. At the same time, the evolution of ethics, concepts of community, and rights, points to the extension of the ethical principle beyond purely human interactions, to include other sentient beings and ecosystems. The types of value we place upon nature, and the extent of recognition we lend to its intrinsic value, will be central to deciding its 'right' to existence. For tourism to be 'well designed' and play a sustainable part in a future green economy, there is a minimum requirement to conserve its resource base and recognize the values of nature beyond simply those that can be economically quantified through market calculations. It is also probable that pressure on land, biodiversity and ecosystems will result in tourism coming into competition with other 'environmentally friendly' industries for land use, as is demonstrated in the case study of Mabira Forest Reserve in Uganda, and which raises an interesting debate over the use of environmental ethics in developmental decision-making.

Activity

Case Study: 'Eco-tourists save forest "jewels" from bulldozers'

So ran a newspaper headline concerning the debate over the future of the Mabira Forest Reserve near Lake Victoria in Uganda, often referred to as a 'wildlife jewel' and a location given protected status in 1932. The reserve contains a rich biodiversity of wildlife with over 300 bird species and 200 types of tree, dozens of them found nowhere else in the world. The grey-cheeked mangabey, a species of monkey listed as endangered, is also found in the forest. Alongside providing a habitat for animal and plant species, the forest fulfils a vital role in the ecosystem functions of the Lake Victoria basin. This includes acting as a water-catchment zone for the rivers and streams that feed the lakes of East Africa that aid agricultural production, carbon sequestration in the region, and provide a rich reserve for Uganda's ecotourism industry. It also has a historic value as one of the last remaining portions of the extensive tropical forest belt of Africa that once stretched from Uganda to the Democratic Republic of Congo. The cultural and spiritual values of the forest are also highly important to the Buganda tribe, who believe the spirits that protect their kingdom reside there.

While the forest is protected as a wildlife reserve, it was threatened in 2007 by a proposal from the Sugar Company of Uganda (SCOUL) to replant the land area with sugar cane for the production of ethanol for biofuels. The scheme would have destroyed the biodiversity of the forest and was met with protest from environmental groups. This resulted in violent clashes with security forces, and three activists were killed. The argument over the future of the Mabira

Forest reserve was made more complex by the fact it already had an existing economic use as an ecotourism attraction. A study conducted by NatureUganda estimated the commercial value of tourism and carbon capture in Mabira at US$316 million a year, whereas growing sugar for biofuel production was found to be worth only US$20 million. The importance of ecotourism for Uganda is underlined by it being the second biggest foreign exchange earner.

The rationale for the refusal of the proposed scheme was, in this case, economic – ecotourism being calculated to be more economically beneficial than biofuel production. But the dependence upon the case for conservation resting on its market-defined economic value raises issues about the relative importance of other types of value attached to nature; for example the cultural and spiritual values held by the Buganda. It also raises issues concerning the importance of the intrinsic right of nature to an existence *independent* of human value. Also, if market conditions change in the future to favour the economic returns of biofuel production over ecotourism, does this make the removal of the rainforest ecosystem for its use as agricultural land, 'right'? The need for this kind of debate about the wider values of nature, beyond the economic, is underlined by the recent re-emergence of the SCOUL proposal.

Sources

Nakkazi, E. (2011) 'Ugandans Mobilise To Save Mabira Forest From Sugarcane Plantation' in *The Ecologist*. Online, available at: www.theecologist.org/how_to_make_a_difference/wildlife/1057616/ugandans_mobilise_to_save_mabira_forest_from_sugarcane_plantation.html (accessed 27 February 2013).

Sapp, M. (2013) 'Revisiting Mabira, Five Years On'. Online, available at: www.sugar-line.com (accessed 20 January 2013).

Questions

1 Ignoring the economic calculations presented in the case study, using the ethical perspectives of 'instrumentalism', 'conservation' and 'deep ecology', present an argument about whether the Mabira rainforest should be used for ecotourism or biofuel production.
2 Based upon the theory of utilitarianism, which development option could be considered 'right' – biofuels or ecotourism?
3 Which argument do you believe is most persuasive for making a decision – the economic one, or that of environmental ethics? Explain your reasoning.

References

Becken, S. (2007) 'Tourists' Perception Of International Air Travel's Impact On The Global Climate And Potential Climate Change Policies' in *Journal of Sustainable Tourism*, 15: 351–68.

Chesshyre, T. (2005) 'Lean, Keen and Green' in *The Times Travel Supplement*, 27 August: 4–5.

Energy Saving Trust (2007) *Green Barometer: Measuring Environmental Attitudes*, London: Energy Saving Trust.

Hares, A., Dickinson, J. and Wilkes, K. (2010) 'Climate Change And The Air Travel Decisions Of UK Tourists' in *Journal of Transport Geography*, 18: 465–73.

Holden, A. (2003) 'In Need Of New Environmental Ethics For Tourism' in *Annals of Tourism Research*, 30: 92–108.

Jamieson, D. (2008) *Ethics And The Environment: An Introduction*, Cambridge: Cambridge University Press.

Leopold, A. (1949) *A Sand County Almanac*, Oxford: Oxford University Press.

Naess, A. (1973) 'The Shallow And The Deep, Long-Range Ecology Movement: A Summary' in *Inquiry*, 16: 95–100.

Nakkazi, E. (2011) 'Ugandans Mobilise To Save Mabira Forest From Sugarcane Plantation' in *The Ecologist*. Online, available at: www.theecologist.org/how_to_make_a_difference/wildlife/1057616/ugandans_mobilise_to_save_mabira_forest_from_sugarcane_plantation.html (accessed 27 February 2013).

Nash, R.F. (1989) *The Rights of Nature: A History of Environmental Ethics*, Wisconsin: The University of Wisconsin Press.

Sapp, M. (2013) 'Revisiting Mabira, Five Years On'. Online, available at: www.sugarline.com (accessed 20 January 2013).

Stone, D.C. (1972) 'Should Trees Have Standing? Towards Legal Rights for Natural Objects' in *Southern California Law Review*, 25: 450–501.

Thoreau, H.D. (1993) *Civil Disobedience and Other Essays*, New York: Dover Publications (reprint of *Walking* (1862)).

Thornton, J. (2007) 'Can Lawyers Save the World?' in *The Ecologist* (June): 38–46.

UNEP (2005) *Integrating Sustainability into Business: A Management Guide for Responsible Tour Operations*. Paris: United Nations Environment Programme.

UNEP (2011) *Green Economy: Pathways to Sustainable Development and Poverty Eradication: A Synthesis for Policy Makers*, United Nations Environment Programme, Nairobi, Kenya.

UNWTO (2012) *UNWTO World Tourism Barometer*. Madrid: World Tourism Organisation.

UNWTO (2013) 'Global Code of Ethics for Tourism'. Online, available at: www.ethics.unwto.org/en/content/global-code-ethics-tourism-article-3 (accessed 26 February 2013).

Vardy, P. and Grosch, P. (1999) *The Puzzle of Ethics*, London: HarperCollins.

WCED (1987) *Our Common Future*, Oxford: Oxford University Press.

Wenz, S.P. (2001) *Environmental Ethics Today*, New York: Oxford University Press.

Zografos, C. and Allcroft, D. (2007) 'The Environmental Values of Potential Ecotourists: A Segmentation Study' in *Journal of Sustainable Tourism*, 15: 44–65.

Part II
Situating the self in ethical consumption

6 A fresh look into tourist consumption

Is there hope for sustainability? An empirical study of Swedish tourists

Adriana Budeanu and Tareq Emtairah

Introduction

Progress towards a more sustainable future of tourism is conditioned by the simultaneous improvements of the production and consumption of leisure services. Consequently, the role that tourists can play in complementing industry efforts to optimize the supply of tourism services has preoccupied tourism researchers for over two decades (Forsyth 1997; Miller 2001; Eligh *et al.* 2002). The particular line of inquiry that follows the research on environmental behavior of tourists shows various results, some positive – suggesting the arrival of a new class of environmentally concerned tourists (Böhler *et al.* 2006) – while others are more circumspect to such claims, doubting the consistency of tourist self-proclaimed interests in environmental aspects of their holidays (Mihalic 2001; Budeanu 2007; Bergin-Seers and Mair 2009), which is rarely reflected in practice by more than niche groups (Chafe 2005).

Based largely on studies of single streams of tourist consumption of resources or services (Hares *et al.* 2010), research results are ambiguous and sometimes contradictory, and provide weak support to designing policies for a sustainable development of tourism. There is a deficiency of studies that contextualize the use of services that are supportive of sustainability goals (e.g. environmentally friendly services) relative to the multitude of options that tourists have to navigate before choosing one service over another. Aiming to fill the gap, this study looks at parallel streams of consumption of different forms for transportation, accommodation and optional leisure, in order to get a comparative perspective on the patterns of tourist choices and their environmental awareness.

This chapter presents results from an investigation that evaluates the factual consumption of three services – transportation, accommodation and optional leisure – by 428 tourists, in two of the most popular Swedish destinations, Gotland Island (East coast to the Baltic Sea) and Åre Ski resort (North West). The local conditions in destinations define the environmental relevance of tourist choices, and therefore they are integrated in this study by the introduction of the term of *environmentally preferred option* (EPO), which distinguishes the service with the least potential to generate negative impacts from all the services available at a specific location. Furthermore, the study evaluates the awareness of tourists about environmental impacts related to the three types of service. By design, the study inquires only about tourists' actual use of services (i.e. choices already made at the time of the study) and factual awareness (i.e. concrete knowledge of impacts), avoiding hypothetical questions in order to minimize respondent bias (Kahneman and Thaler 2006). The first part of the chapter includes an overview of theoretical considerations that frame current knowledge about tourist environmental behavior, followed by methodological considerations that guided the study and a brief presentation of results. In the second part, the comparative evaluation of parallel streams of consumption and the significance of environmental awareness are discussed and relativized in the context of tourist consumption.

Sustainable tourism and the role of consumers

A sustainable development of tourism is institutionalized as a guiding principle for economic development that does not come at the expense of human and natural ones, and is concentrated around concepts such as futurity, equity and holism (Saarinen 2006). The practical application of this principle consists of raising efficiency and effectiveness of tourism provision, through recycling, energy efficiency and the optimization of value creation processes (e.g. closed loop management), accompanied by a redefinition of tourist consumption of natural and cultural resources at levels that create meaningful experiences within the limits of available resources. In the accommodation sector, for example, eco-efficiency measures are becoming popular (Bohdanowicz and Zientara 2008; Accor Group 2007), while transport and mobility sectors take a leading role in technological and infrastructure eco-innovations (Gronau and Kagermeier 2007; Gössling 2009). Along the same lines, destinations are starting to implement sustainable development strategies (UNWTO 2004; Schianetz *et al.* 2009) and some tour operators take the lead in stimulating the adoption of sustainable practices along tourism supply chains (Budeanu 2009) and in destination networks (Schianetz *et al.* 2009, Budeanu 2009; Ferhan 2006; Hudson and Miller 2005). However, maintaining a good quality of natural and cultural tourism resources is complicated by the low interest of tourists in purchasing sustainable tourism holidays or services such as eco-labeled accommodation (Budeanu 2007; Chafe 2005).

Almost two decades ago, the arrival of a "new tourist" was advocated with enthusiasm (Poon 1994) together with hopes that it would be accompanied by high commitments to environmental and social responsibilities. The belief that the number of tourists with high environmental values is increasing (Sharpley 2001) was perpetuated further by a wave of energy related to the celebration of 2002 as the International Year of Eco-Tourism. Subsequently, a lot of evidence started to accumulate about the high environmental and social awareness of tourists (Böhler *et al.* 2006, Fairweather *et al.* 2005, Jensen *et al.* 2004, Baysan 2001) and are willing to pay for environmentally friendly holidays (Hudson and

A fresh look into tourist consumption 85

Ritchie 2001; TripAdvisor 2011) But other authors question the existence of a sustainable tourist (Peattie 2001). Research shows a remarkable inconsistency between environmental awareness and actual behavior, with less than 1 percent of tourists actually behaving in an environmentally friendly way (Sharpley 2001). High percentages of visitors with positive feelings towards nature still travel by car (Barr *et al.* 2010) and records show that only 50 percent of German tourists translate into practice their self-declared environmental awareness (Mihalic 2001). Such conflicting research results indicate that the evidence of the alleged environmental tourist is rather weak, and does not enable an accurate evaluation of the actual tourist behavior (Budeanu 2007; Hares *et al.* 2010, McKercher *et al.* 2010; Dickinson and Robbins 2008).

The environmental behavior of tourists

Some early studies identify and classify tourist behaviors into groups labeled as "soft green" (Swarbrooke and Horner 2001) or "very dark green" (Hudson and Ritchie 2001), which suggests they also have a predisposition to behave in line with sustainability principles. In other studies, the sustainable behavior of tourists is derived from their interests in environmentally relevant activities (Kaae 2001), their environmental convictions (Hudson and Ritchie 2001; Swarbrooke and Horner 2001), or specific activities such as their use of transport during holidays (Böhler *et al.* 2006, Götz *et al.* 2002). Such studies contribute to the understanding of various tourist environmental behaviors; however, their input is limited by the fact that they isolate tourist activities from the destination where they take place, when in fact it is the local context which determines whether tourist consumption of resources is a negative impact of tourism or not. Overlooking the contextual situation of tourist activities limits the implications of such studies for policy and practice.

Besides concerns, tourists are also credited with holding an environmental behavior when they display a high willingness to pay to compensate for the eventual damages generated by their activities at their various destinations (UK Department for Transport 2007). Frequently, companies such as KLM, SAS or Thai Airways praise the "environmental behavior" of customers who agree to pay for offsetting the CO_2 emissions from flying to and from their destinations. Similarly, when tourists display an interest in maintaining the quality of nature in destinations as part of their tourist product (hiking, rafting, skiing) they are qualified as "environmental tourists" and expected to use environmentally friendly products or services throughout their holiday (Hudson and Ritchie 2001). Although tempting for political or reputational reasons, generalizations based on singular acts such as paying for CO_2 compensations are not representative, and should not be used as equivalents for tourist behaviors. Research shows that people are increasingly interested in the environmental quality of destinations (European Commission 1998), with "unspoiled nature" becoming a highly important criterion when choosing holiday destinations (Ayala 1996; Mustonen 2003). Ideally, an environmental tourist would also choose environmentally friendly services (transport, hotels, food), demonstrate a respectful attitude to locals, and avoid performing activities that could harm nature and communities. However, little is known about whether such implied concerns are actually reflected in their daily holiday routines. Furthermore, evidence shows that, driven by curiosity, and often ignorant of consequences, eco-tourists can easily disturb the biodiversity and equilibrium of these same ecosystems that are the subject of their attentions (Grossberg et al. 2003). Such acts contest the value of preferences for nature-related holidays as reliable proxies of tourist environmental behavior (Mustonen 2003).

Environmental awareness is the most cited indicator of tourist environmental behavior. On numerous occasions tourists are found to be aware of the negative consequences of tourism (Mihalic 2001; Hudson and Ritchie 2001; Khan 2003; Wurzinger and Johansson 2006) and about tourism eco-labels (Fairweather et al. 2005). There are also indications that tourist awareness about the environmental and social aspects of their holidays is increasing (Kang, Moscardo 2006). Often examined in this context are the kind of environmental values, beliefs or attitudes which might indicate the potential environmental behavior of tourists. (Fairweather et al. 2005) However, both attitudes and values are often found to have less significance in determining a behavioral change (Dickinson and Dickinson 2006; Reiser and Simmons 2005), especially in habitual situations (Garvill et al. 2003). While awareness and values are important prerequisites of environmental behavior, it is clear that their predictive power is limited (CREM 2000; Wurzinger and Johansson 2006).

Characterizing behavior by looking at proxies such as values, attitudes or a willingness to pay, all of which are hypothetical in nature, leads to a rather ambiguous representation of what the sustainable or environmental behavior of tourists may be. The lack of precision about the types of tourist behavior that can contribute to the sustainable development of tourism brings only a partial understanding of its complex mechanisms, allowing easy generalizations that serve to perpetuate the inefficiencies of policy intervention. More accurate investigations of tourist environmental behavior (Budeanu 2007; Garvill et al. 2003) could enable a better positioning of corporate and policy pursuits towards sustainable tourism. This chapter presents results from an investigation that took as its point of departure the assumption that more clarity regarding tourist choices will facilitate the consolidation of knowledge that (now) describes fractional aspects of sustainable tourist behavior.

Methodology

Respondents and sampling

The data for this study was collected through two series of interviews conducted in two of the most popular destinations of Sweden – Åre (a ski destination) and

Gotland (a beach destination) – covering the peak season in each destination. The interviews took between 15 and 20 minutes, and answers were recorded on predesigned paper forms (to reduce recording time).

Falling into the pragmatic qualitative research paradigm, the findings of this study do not enable a wide generalization of results. Rather, the aim was to explore traveling experiences in detail, in order to identify patterns of service consumption and associated environmental awareness. The results are meant to provide empirical grounding for developing theoretical approaches to sustainable consumption in tourism. With this in mind, the representativeness of the sample of respondents was less significant than the detail of the traveling accounts. However, the wide range of locations (minimum ten in each destination) and the unbiased selection of respondents (only children and residents were excluded) give sufficient support to claim that the results represent the average use of services for those respective groups of winter and beach tourists. Potential respondents were approached individually (even if they were part of a group) and their answers constitute distinct inputs in the analysis. On the assumption that respondents might be more willing to respond to questions if they were on a break from their regular holiday activities, tourists were approached in rest areas (e.g. bus stops, promenade areas, queues for ski-lifts, parks or town squares, etc.). Valid responses from 100 interviews in Åre and 328 interviews in Gotland were recorded and analyzed as two separate data-sets using the SPSS v 18.0 software.

Tables 6.1 and 6.2 give an overview of the socio-demographics of respondents in the survey sample as well as the trip characteristics. Table 6.1 shows key demographic characteristics of the sample, with a dominance of high income, educated tourists in both destinations. The proportion of female respondents is

Table 6.1 Number and percentage of respondents in Åre and Gotland, by socio-demographic category

		Åre (N=100)		Gotland (N=328)	
		N	%	N	%
Age	20 or less	8	8	45	13.8
	21–30	26	26	142	43.6
	31–49	48	48	96	29.4
	50 and over	18	18	43	13.2
Gender	Male	56	56	113	34.5
	Female	44	44	215	65.5
Education	Primary-level	6	6	1	0.3
	Secondary-level	28	28	117	37.8
	University-level	66	66	191	61.8
Income	Low	34	34	74	33.5
	Medium	48	49	108	48.9
	High	16	16	39	17.6

Table 6.2 Trip characteristics

	Åre (N = 100) %	Gotland (N = 328) %
Traveling party		
Alone	0	7.8
Couples	10	19.5
Family with children	56	22.1
Mixed group	34	50.6
Place of origin		
Norrland (northern part of Sweden), Norway, Finland, Denmark	7	5.2
Svealand	62	59.5
Götaland	27	22.6
Outside Nordic countries	4	12.8
Experience with destination		
First time	21	25.5
Been here before	79	74.5

significantly higher in Gotland than in Åre. There have been previous observations of a higher propensity among women towards participation in on-the-spot interviews (Emtairah *et al.* 2012), which could explain the higher proportion found in Gotland, but not in Åre.

The two samples differ in terms of the predominant age profile of visitors, Gotland standing out with over 43 percent of visitors being under 30 years old. These findings are reflected in the structure of the traveling party – in Åre vistors questioned were predominantly from families with children (56 percent), while in Gotland half of respondents were with a group of friends. The results fall in line with the profile of the two destinations, with Åre being known as a family destination, and Gotland as a "youth" destination. While Gotland is largely preferred by people from the neighboring regions (Svealand and Götaland), tourists visiting Åre come from far away. For both destinations nearly two-thirds of tourists were repeat visitors.

Measures

Questionnaire design

This investigation aims to identify patterns in the tourist consumption of services and the environmental awareness associated with it. A key aspect that guided the research design was the intention to minimize respondent bias (Kahneman 2003), a common weakness of research related to environmental issues (Huesemann 2002). Consequently, the questionnaire design had a progressive approach, leaving questions directly associated with the environmental theme to the very end. The questionnaire was tested and modified twice before use in the field. Additional changes were made to fit the type of destination and holidays investigated. Questions about tourist leisure activities were different for the two destinations, as one is a winter resort and the other a summer destination.

The questions were divided into three sections. One set of questions focused on background information about the holiday (length of stay, origin of travel, traveling party, etc.) and socio-demographic data about the respondents (age, education, income, etc.). Another section was dedicated to mapping out the tourist consumption of services, evaluated in terms of their choice of accommodation, transport and optional leisure pursuits. In this study, "optional leisure" is a term covering leisure activities that are additional to the primary holiday activity (which is skiing in Åre and sunbathing for Gotland). The last section was dedicated to identifying tourists' factual knowledge about environmental impacts related to the holiday services they were using (transport, accommodation, leisure), which serves as a proxy for their environmental awareness.

Tourist consumption

For this investigation, tourist behavior is limited to consumptive behavior, understood as the sum of choices related to their selection of holiday services. The deconstruction of holidays by the potential of different service elements to generate environmental impacts (Budeanu 2007) led to the selection of transportation, accommodation and leisure activities as those services most likely to contribute to tourism impacts. Traditionally, the emphasis in environmental research is given to holiday choices made prior to arriving at a destination, and in particular the choice of destination or the mode of transport used to get there and back. However, estimates show that 25 percent of CO_2 emissions generated by tourism come as a result of accommodation and leisure choices (UNEP, UNWTO 2008), a percentage that justifies including them in this investigation. In order to minimize respondent bias, the study did not inquire about hypothetical situations, relying instead only on performed behavior (choices already made at the time of the interview). Moreover, the research did not investigate the causal relationships between traveling motivation and choice, but was limited only to the identification of patterns of choice. However, as subjects were asked to give an account of their choices, the rate of response is not always 100 percent.

Environmental awareness

Despite its weakness as a descriptor of behavior, awareness is a key condition that precedes responsible activities, and its role is important to investigate. Factual awareness was identified by asking people to specify one or more environmental impacts they knew of that was related to each of the three holiday services that were investigated (travel, accommodation and optional leisure activities). Asking respondents to demonstrate factual knowledge about the environmental impacts of holiday services they have actually used serves to eliminate hypothetical situations and significantly reduce bias (Kahneman and Thaler 2006). Respondents who named one or two impacts were considered to have an average awareness, while those ones who could name three impacts (or more) were considered to have high awareness.

Environmental significance of service consumption

The holiday experience is the cumulative result of multiple successive decisions (Jeng and Fesenmaier 2002) and therefore a better understanding of holiday behavior may be obtained by examining these choices in parallel. At present, the multi-dimensional character of tourist decisions and choices is largely overlooked by research (Jeng and Fesenmaier 2002; Garvill *et al.* 2003). Therefore, this investigation distinguishes the consumption of substitutable services that satisfy the same function (Table 6.3), in order to unveil similarities and differences, and to enable a better understanding of their relative environmental significance.

This research introduces the term *environmentally preferred option* (EPO) to distinguish the service version with the least potential to generate negative impacts from all the available services that serve the same function. The designation of a service as being an EPO is based on its relative impact-intensity per person (tourist) compared with its alternatives, in the context of a specific location. Hence, the same service can be an EPO in one location, but not in another. A common example of an EPO is public transportation, which is not only the least impact-intensive of all transportation modes (Böhler *et al.* 2006) it is also readily accessible almost everywhere. In remote destinations, however, where public transport often runs dry, the inefficiency of its use makes it unsuitable to be considered an EPO.

Table 6.3 Options for transportation, accommodation and leisure

Åre (ski destination)	Gotland (beach destination)
Transportation	
Train and/or bus (EPO)	Train and/or bus (EPO)
Plane	Plane
Car	Car
	Boat
	Other
Accommodation	
Hotels and hostels (EPO)	Hotels and hostels (EPO)
Apartments and cottages	Apartments and cottages
Private accommodation	Private accommodation
	Other (e.g. camping, boat, trailers)
Optional leisure activities	
Energy intensive activities	Energy intensive activities
Less energy intensive activities (EPO)	Less energy intensive activities (EPO)

For accommodation, the selection of EPOs differs depending on the location, due to specific measures that destinations can take to make tourist accommodation more environmentally friendly. Ideally, location-specific impact evaluations should be used for identifying EPOs for tourist services, but such studies are rare. This investigation used a comparative study of environmental impacts of tourist accommodation in Finland (Salo et al. 2008), a country with similar climate and infrastructure characteristics to Sweden. In the absence of similar research into Swedish hotels, and based on the similarities between the Finnish and Swedish climates, this study takes the Finish research as a benchmark for energy efficiency in accommodation. According to the Finnish results (Table 6.4) large hotels with higher occupancy levels use fewer resources in terms of energy/person/night, and can therefore be considered to be the EPO for accommodation in this study.

For leisure activities, comparative studies of environmental impacts are rare, and in this case Australian research (Becken 2001) was used as a reference because it has the most comprehensive overview of energy consumption by various activities. For the Swedish investigation, leisure activities were aggregated (Table 6.5) based on Becken's classification (2001) into "energy intensive" activities (all above 200 MJ/tourist) and "less energy intensive activities" (under 200 MJ/tourist).

Collecting data from destinations with very different profiles – beach, ski – was carried out in order to identify the existence of any significant differences between the patterns of service consumption between winter tourists and beach

Table 6.4 Natural resource input per service unit for accommodation (overnight)

Type of accommodation	Total material resources (including abiotic: buildings, electricity and heating; biotic: building and furniture; water) (kg)	Air resources (electricity, heating) (kg)
Radisson SAS Seaside Hotel	37	14
Sokos Hotel Arina	45	18
Simple cottage (29 m^2)	50	14
Well equipped "cottage" (140 m^2)	125	27

Source: Salo et al. (2008).

Table 6.5 Categories of leisure activities based on energy intensity

Energy intensive leisure activities	Heli-skiing, scenic flights, diving and boat activities, jet skiing, windsurfing/kite-surfing, snowmobiling, sauna/pool activities, etc.
Less energy intensive leisure activities	Visiting museums and historical sites, beach activities, fishing, guided walks, hiking/bird-watching/visiting caves, rafting, visiting museums and visitor centers, kayaking, biking, horseback riding, ice skating, dog sledding, etc.

(summer) tourists. The same questionnaire was used (with only small adjustments to fit the local contexts) in both the ski resort (Åre, in Northern Sweden), and at the beach destination (Gotland, Sweden). A comparison of findings between the two locations enabled the authors to reflect on the possible implications for Swedish tourist destination managers.

Frequency analyzes have been carried out to establish background characteristics including the socio-demographics of respondents, the types and percentage of leisure product choices, levels of awareness, and other variables included in the questionnaire. Exploratory analyzes have also been made in order to examine relationships between service choices (transport, accommodation and leisure) and the different independent variables using bivariate analyses (cross-tabulations). Person Chi-square tests with significance level 0.05 were used to determine significance in the relationships. Most of the Chi-square tests had to be abandoned due to the small cell counts, i.e. small number of cases in more than 20 percent of cells in the cross tabulation tables.

Results

This investigation has the purpose of mapping out the tourist consumption of three holiday services: transportation, accommodation and optional leisure activities. The investigation searched for patterns in the use of holiday services and patterns of awareness, by comparing different streams of consumption in two different samples: beach tourists and ski tourists. The use of services with less harmful potential to the environment is compared to the use of services that have a high potential to contribute to tourism impacts. Last of all, relationships between awareness and choice are explored.

Service consumption use

The results show the car to be the preferred mode of transport to Åre (47 percent) (Table 6.6). However, public transport is nearly as important (43 percent) and is also the dominant transport mode for traveling to Gotland (43.9 percent). The high popularity of this mode of travel among tourists is not surprising considering that Sweden is a country renowned for its effective promotion of public transportation. Slightly atypical, in fact, is the high usage of cars for travel to Åre, a destination where the well-organized public transportation system to, from, and within, is a very convenient option. In Gotland, where public transportation is less convenient and involves a mix of different modes (train, bus and ferry), it is still used by a high percentage of tourists. These findings suggest that the importance of "convenience" as a motivating factor for transport choice may be declining. The share of air transportation users are significant, though minimal, to both destinations.

The most popularly form of accommodation in Åre are rented apartments or cottages, with more than two-thirds of tourists using this type of accommodation. In Gotland only about half these numbers use this type of accommodation.

Table 6.6 Tourist choice of travel, accommodation and optional leisure

	Åre (N=100)	Gotland (N=328)
	(%)	(%)
Travel mode		
Train and/or bus	43	43.9
Plane	10	15.9
Car	47	33.0
Boat	n/a	5.6
Other	n/a	1.6
Accommodation		
Hotels and hostels	9	17.1
Apartments and cottages	68	35.5
Private accommodation	21	26.6
Other (e.g. camping, boat, trailers)	0	21.1
Leisure activities		
Energy intensive activities	5	7.9
Less energy intensive activities	24	96.6

The next most popular category is private accommodation (secondary homes, staying with friends and family), which has a comparable share in both samples (21 percent in Åre and nearly 27 percent in Gotland). The third most popular accommodation in both destinations are hotels and hostels. The high popularity of apartments among tourists in Åre is explained by the reduced variety of options compared to Gotland. As a summer destination, the geo-climatic conditions in Gotland enable the use of various alternative accommodations such as tents, camping trailers and private boats. They are aggregated as the "other" category in the accommodation choices for Gotland, and cover a large share of rentals (21 percent). In Åre many of the tourists are "loyal," and have developed a strong attachment to the destination and have long-term relationships with the owners of cottages and apartments.

Optional leisure makes for the most diverse stream of consumption in the two destinations. At the same time, most of the optional leisure activities are low consumers of energy, and count as environmentally preferred options for optional leisure. In this respect the results are noticeable in Gotland, where 96 percent of tourists participate in (in addition to the regular activity of sunbathing) less energy intensive leisure activities. As the total percentage of leisure activities is not cumulative, there are instances where respondents identified activities in both categories.

Environmental awareness

With regard to factual environmental awareness, respondents were asked to name one or more impacts associated with each product. The answers were coded into a scale starting with "none" (indicating low awareness), one or two impacts (indicating medium awareness), and three or more impacts (indicating high awareness). Table 6.7 gives an overview of levels of awareness of various impacts (by percentage) of respondents for each of the three service product choices. Overall, the results show that the share of tourists with a medium awareness about environmental impacts are high for all three service areas investigated in both destinations.

Tourists' awareness of transportation impacts has a normal distribution, with a distinct majority falling into the category of "medium" awareness, flanked on either side by comparable shares of "ignorant" and "highly aware" tourists. For accommodation, tourists with medium awareness clearly dominate in Åre but have comparable shares with the "ignorant" group in Gotland. The same pattern is visible for the users of optional leisure. Overall, Gotland has high percentages of "ignorant" tourists who could not mention any impacts generated by accommodation (48 percent) or leisure activities (62 percent), when compared to Åre. In terms of awareness of leisure impacts, and comparing the two destinations, the analysis shows that a higher percentage of respondents in Åre compared to Gotland fall into the medium to high awareness categories (75 percent in Åre and 38 percent in Gotland). In this context, there seems to be a clear sense of energy intensive activities compared to less energy intensive activities (e.g. snowmobiles versus dog sledding) in destinations like Åre, while the energy intensive activities are not as obvious in Gotland (e.g. shopping versus cultural sightseeing).

Environmental awareness and choice relationships

The exploratory analyzes on the relationships between the choice of service (transport, accommodation and optional leisure) and the different independent variables using bivariate analyzes (cross-tabulations), confirm only two

Table 6.7 Environmental awareness

	Åre (N=100)	Gotland (N=328)
	%	%
Awareness of transport impacts		
None	18	16.1
1 or 2 impacts	68	72.7
3 or more impacts	14	11.2
Awareness of accommodation impacts		
None	34	48,1
1 or 2 impacts	47	43.7
3 or more impacts	19	8.2
Awareness of activity impacts		
None	24	62.3
1 or 2 impacts	60	34.0
3 or more impacts	15	3.7

significant relationships with the traveling party and the distance from home. Most of the Chi-square tests used to determine significance in the relationships had to be abandoned due to the small cell counts, i.e. small number of cases in more than 20 percent of cells in the cross-tabulation tables. Similarly, the frequency analyzes performed between environmental awareness and the selection of environmentally desirable services show no significant relationship.

Discussion of environmental implications

Aiming to provide a comprehensive picture of tourist consumption of holiday services, this study mapped out the parallel use of transportation, accommodation and optional leisure. Based on the findings, this section discusses the environmental significance of identified patterns of consumption relative to each other. In order to distinguish services that are most environmentally desirable from a destination perspective, and to contextualize service consumption, the study introduces the term of *environmentally preferred option* (EPO). The lack of contextualization has been cited as an important limitation for the applicability of studies of tourist environmental behavior in – for example – policy development (Hjalager 1999). The concept of EPOs moves away from the narrow focus on characterizing the individual user, or choices, and opens up possibilities to discuss an array of influences that converge into final decisions: the availability of services, the weighing of advantages and disadvantages of specific choices for the user and for the destination. In contrast with the use of single benchmarks across multiple geographical areas, the EPO concept acknowledges and incorporates the diversity of local natural and manmade conditions, and which can be connected to policy interventions to help evaluate their effectiveness.

A good use of EPOs

The main goal of this study is to identify patterns for substitutable holiday services and explore the relevance of specific environmental awareness for the choices made. Overall, the findings show that the environmentally preferred options (EPOs) for transportation, accommodation and optional leisure are being used extensively by both ski and beach tourists in Sweden. The results confirm earlier studies into individual services such as transportation, for example (Böhler *et al.* 2006). By using the EPO concept, this study emphasizes also the role of the context in tourist choices. Looking at the data-sets from both destinations it is obvious that Åre has distinctive clusters of service users, while in Gotland the users are widely distributed. One reason for this result might be the significantly larger area of Gotland covered by this study (investigations were carried out in eight different locations on the island) which enabled access to more diverse users when compared to the investigations completed in the one locality of the ski resort of Åre. At the same time, more types of services are available in Gotland than in Åre, which is reflected in the user patterns. More interesting distinctions can be identified when looking at the three streams of consumption, as discussed below.

For transportation, the train and bus have a consistent share of around 43 percent of all users in both destinations, which confirms findings from earlier research that indicates the adoption of public transport in Sweden is high. Although investigating the motivating factors behind its use was not within the scope of this study, some possible explanations became evident during the analysis, and are briefly presented below. One important explanation for the appeal of public transport to tourists might be the consistent effort of transport operators to create enticing offers aimed at them. However, the frequency analysis in this study did not confirm the relationship between transport choice and price. Instead, there is a significant association between the choice of transport and the composition of the traveling party ($x2(1)=25.36, p=0.001$). This result indicates that the liklihood of people choosing to use public transport are 6.65 times greater if they are single (whether traveling alone or in groups) than if they travel as a family. Furthermore, the recording of similar shares for car users and public transport users in Åre suggests that price incentives for environmentally friendly transportation may have reached the limit of their persuasive potential. In order to provide systematic explanations for these patterns of choice it is necessary to investigate in more detail the effect of tourist-specific and non-tourist-specific motivators (such as the experiential value of different services, or the price of petrol (which makes the car use less desirable)).

Positive results are also found for optional leisure, where a large majority (96 percent) of tourists in Gotland and almost a quarter of the skiers in Åre use less energy intensive activities aside from their primary leisure activity (beach activities and skiing). The result may reflect the dominant presence of Scandinavian tourists, who are known to prefer nature-based leisure activities such as hiking and biking (Zilinger 2005). Furthermore, the climatic conditions of the beach destination enable more diverse outdoor activities and explain their extensive consumption. A lot of this success can be attributed to the commonly agreed social practice in Scandinavian countries of valuing and promoting outdoor activities.

For accommodation, the findings show that large establishments (chosen as EPOs for this study) were the least popular with both the skiers in Åre and the sun tourists in Gotland; in contrast with the small-scale dwellings (apartments or cottages) were popular. One reason for this situation may be the low availability of large establishments. For example, in Åre, the four largest hotels have a total capacity of nearly 500 rooms, insufficient for the influx of 50,000 visitors a year. Although they are among the destinations with the highest concentration of visitors in Sweden, Åre and Gotland are not typical mass tourism destinations, and the dominant form of accommodation is small-scale dwellings. Furthermore, both destinations have a large number of secondary homes, which explains the fact that nearly a quarter of visitors use private accommodation. In this context, it seems that the structure of the existing accommodation sector has locked in the usage patterns for accommodation services.

Tourists can contribute a lot to the sustainability of a destination by selecting their accommodation based on environmental considerations, among other criteria (Salo *et al.* 2008). However, they may have difficulties making the selection themselves in the absence of sufficient information about the available options and their likely environmental impact. In Åre, for example, Hotel Diplomat Åregården is the only establishment environmentally certified with the Nordic Swan Ecolabel, but information about this status is difficult to find. The same can be noted about Gotland, where information about environmental accommodation (as well as other tourist services) is sketchy. Although offering information about the environmental performance of different services may not change dramatically tourists' choices, it might tilt the balance in favor of such choices.

Looking across all three consumption domains – transport, accommodation and optional leisure – the champion sector is transportation, where almost half of users in both destinations adopted EPOs, followed closely by optional leisure activities, where the EPO has the highest figure in the entire sample. These results represent positive feedback for destination management organizations in Åre and Gotland, confirming the success of ongoing initiatives and campaigns for the adoption of public transport and outdoor leisure activities. More effort could be put into making environmentally friendly accommodation more conspicuous. However, since accommodation is largely centered in the private sector, any intervention by destination management organizations requires public-private partnerships and collaborative initiatives.

The results are encouraging and give significant impetus for further research. Special attention in tourism studies is given to the influence of enabling factors such as the type of destination and attractions, the length of trip, the degree of familiarity with the destination, and the available resources such as money, time and knowledge (Pizam and Mansfeld 1999). By focusing on the explanations of specific choices, such studies overlook the fact that holidays involve a plurality of decisions, triggered by different factors (Hyde and Laesser 2009) and with diverse environmental consequences. Further research into the chains of causality that lie behind parallel streams of consumption could unveil possible deterministic patterns and identify common paths to shifting tourist behavior.

Evaluating awareness

The second scope of this study, the investigation of tourists' awareness of the impacts of different services, shows less unified patterns, with relatively higher levels of environmental awareness about transportation impacts when compared to awareness of accommodation and leisure impacts. These results confirm the findings of previous studies where awareness levels about transportation were high (Böhler *et al.* 2006). But due to different methodologies used, comparisons should be made with caution. The results of this study do not confirm the supposed relationship between environmental awareness and service choice (Kang and Moscardo 2006), confirming emerging skepticism about the predictive strength of awareness (Dickinson and Dickinson 2006).

Of all three services investigated, transportation is a champion in terms of awareness, as well as use. For instance, 68 percent of respondents in Åre, and close to 73 percent from Gotland, could identify one or two impacts associated with transport. This figure drops to an average of 45 percent for both destinations when it comes to awareness of the impacts from accommodation choice. The cumulative share of medium and highly aware tourists are high for users of all three services, with a minimum of 38 percent for users of optional leisure (Gotland) and a maximum of 84 percent for transportation users (Gotland). These findings are the effect of the historically high exposure of Swedish society to environmental information. One can easily find a parallel in the emergence of the public environmental debate on topics such as nature preservation and individual responsibility for maintaining natural assets. In Scandinavian countries, climate change is a common topic for discussion in the media and in public debates. Consequently, it might have been easier (or quicker) for respondents to identify impacts from transport than, for instance, impacts from using accommodation (construction impacts, energy use).

At the same time the awareness about accommodation and leisure shows an interesting pattern, with Åre having a share of medium aware tourists similar in size to the share of "ignorant" tourists in Gotland who could not mention any impact from accommodation or beach-related leisure. The difference can be explained by considering the characteristics of the dominant customer group. While younger tourists (predominant in Gotland) are commonly seen as having fewer environmental concerns, (Pizam and Mansfeld 1999) families (which are predominant in Åre) have higher levels of concern (Wurzinger 2003).

Overall the results demonstrate a relatively high environmental awareness when aggregated across all three services investigated, which shows that tourists have a disposition towards understanding issues relating to the impacts of their tourism choices, and maybe makes them able to act in an environmentally responsible manner. Admittedly this result refers primarily to a cognitive awareness (of concrete facts about impacts) rooted in a long tradition of informative transparency characteristic to the Scandinavian public domain. Despite the absence of a confirmed direct relationship, the recorded high levels of awareness, and the predominant use of the EPO for two of the three services (transport and leisure) gives sufficient grounds for two conclusions. First, it is clear that tourists in Sweden understand complex information about environmental impacts such as climate change, but do not relate them to their own holiday activities. Consequently they are missing the motivation to minimize their impacts. The absence of a common discourse on sustainable holidays at sector and policy level is likely to perpetuate the absence of holiday-specific awareness, and implicitly the lack of corrective action. Without personal motivation or external encouragement, tourists cannot be expected to switch to environmentally friendly transportation, especially if they are satisfied with current options.

Second, the distinctive success of public transport in terms of use as well as environmental awareness, may indicate that people are willing to accept news and information about the impacts of other services: such as accommodation. The results suggest that it may be a good time to build on existing awareness (about climate change and the innate Scandinavian value placed on nature) and construct public discourses that include more details about other sustainable holiday services. Otherwise, holiday decisions are too numerous, and creating separate discourses for each choice would be an inefficient use of resources.

The third conclusion is that tourists are more willing to discuss global problems. Historically, both public and private operators in the tourism industry have been reluctant to spoil tourists' relaxation by mentioning environmental concerns or explaining what it means to be socially responsible (Budeanu 2007). Observations made during this study demonstrate that tourists are not disturbed or embarrassed to discuss openly the environmental impacts that their holiday choices have. On the contrary, they were keen to participate in the study, offered a number of suggestions for improving the sustainability of Åre and Gotland, and (some) asked to be kept informed about the results of this study, all of which suggests they took pride from being part of such a dialog. The traditional resistance of industry actors to introduce environmental information in their promotional campaigns may be misplaced; the results of this study suggest that making the environmental performance of their services explicit may resonate with customers and may indeed reflect a shift in public perception about the sustainability of holidays and tourism.

Several methodological choices made during the design of this study reduced the possibility for investigating the motivating factors influencing tourist choice of holiday services. Consequently, the investigation does not aim to provide explanations for tourist choice patterns or predict future behaviors. Instead, this research provides an environmental segmentation of tourists as service users and an overview of their service-specific environmental awareness. For a stronger predictive capability, these results could be complemented by further investigations into the specific reasons for, and barriers to, choosing EPOs.

Conclusion

Aiming to provide a contextualized overview of tourist consumption of holiday services, this study looked at three streams of consumption in parallel: transportation, accommodation and optional leisure. The results show that EPOs have relatively high shares of consumption compared to conventional services. In addition to its empirical content, this research makes a conceptual contribution by suggesting that tourist choices must be understood in the context in which they are being made, through the concept of *environmentally preferable options* (EPOs). Due to its narrow objective, this investigation has only a limited depth into the motivational fabric of tourist choices, a methodological limitation which restricts the possibility of drawing conclusions about the motivations behind consumption patterns. However these initial findings give sufficient grounds for the further testing of possible correlations between specific awareness and specific choice of services, and the spill-over influences of awareness between the three consumption domains.

The large discrepancy in awareness surrounding the environmental impacts of different tourism products shows the absence of a common discourse on sustainable holidays. Despite a relatively high environmental awareness, the impacts remain abstract to the tourists and unrelated to their own choices. Taking advantage of tourists' relatively high level of knowledge about global environmental problems, it may be a good opportunity to create explicit environmental heuristics about sustainable tourism in order to trigger and justify the tourists' engagement in activities that minimize tourism impacts.

Activity

Make groups in class of four or five students, and supply access to a computer that can be connected to the internet. Each group is given a number (1, 2, 3, etc.), and has the task of working together to write a script for a short play about "A day in the life of a sustainable tourist." Students may use examples and illustrations from the internet, audio-visual materials, testimonies from friends and family, input from friends on Facebook, etc. At the end of the script, each group should identify five criteria that reflect why the day qualifies as "sustainable" and prepare one indicator for each criterion (five indicators in total).

Each group should present their script (or act it) in front of the class, explaining their criteria and indicators.

Through an open discussion, facilitated by the teacher, the class will grade each script (between 1 and 5), with 1 being "least sustainable" and 5 being "most sustainable."

Which scenario is most successful?

Suggested reading

Dickinson, J.E. and Dickinson, J.A. (2006) "Local Transport and Social Representations: Challenging the Assumptions for Sustainable Tourism" in *Journal of Sustainable Tourism*, 14 (2): 192–208.

Hughes, G. (2004) "Tourism, Sustainability, and Social Theory" in A. Lew, C.M. Hall and A.M. Williams (eds.), *A Companion To Tourism*, Malden, MA, USA: Blackwell Publishing Ltd.: 498–509.

Cohen, S.A., Higham, J.E.S. and Cavaliere, C.T. (2011) "Binge Flying: Behavioural Addiction And Climate Change" in *Annals of Tourism Research*, 38 (3): 1070–89.

Acknowledgments

The Swedish Research Council FORMAS is gratefully acknowledged for its support to this research as a part of the Sustainable Lifestyles project. The authors thank the municipality of Åre, the management of Skistar Resort in Åre, and the municipality of Visby in Gotland for their time and support during this study. The authors would like to thank all the interviewees for their input and to the masters students from the IIIEE/Lund University for their help in collecting the data in this investigation. Lastly, the authors are grateful to the editors and the reviewers for their constructive comments.

References

Accor Group (2007) *Environment Guide for Hotel Managers*, ACCOR.

Ayala, H. (1996) "Resort Ecotourism: A Paradigm for the 21st Century" in *The Cornell Hotel and Restaurant Administration Quarterly*, 37 (5): 46–53.

Barr, S., Shaw, G., Coles, T. and Prillwitz, J. (2010), "A Holiday is a Holiday: Practicing Sustainability, Home and Away" in *Journal of Transport Geography*, 18(3): 474–81.

Baysan, S. (2001) "Perceptions of the Environmental Impacts of Tourism: A Comparative Study of the Attitudes of German, Russian and Turkish Tourists in Kemer, Antalya" in *Tourism Geographies*, 3 (2): 218–35.

Becken, S. (2001) "Energy Consumption of Tourist Attractions and Activities in New Zealand," Lincoln University. Online, available at: www.lincoln.ac.nz/PageFiles/6834/892_encnsmtn_s3347.pdf (accessed November 10, 2012).

Bergin-Seers, S. and Mair, J. (2009) "Emerging Green Tourists in Australia: Their Behaviours and Attitudes" in *Tourism & Hospitality Research*, 9(2): 109–19.

Bohdanowicz, P. and Zientara, P. (2008) "Corporate Social Responsibility in Hospitality: Issues and Implications. A Case study of Scandic" in *Scandinavian Journal of Hospitality and Tourism*, 8 (4): 271–93.

Böhler, S., Grischkat, S., Haustein, S. and Hunecke, M. (2006) "Encouraging Environmentally Sustainable Holiday Travel" in *Transportation Research Part A: Policy and Practice*, 40 (8): 652–70.

Budeanu, A. (2007) "Sustainable Tourist Behaviour: A Discussion of Opportunities for Change" in *Journal of Consumer Studies*, 31 (5): 499–508.

Budeanu, A. (2009) "Environmental Supply Chain Management in Tourism: The Case of Large Tour Operators" in *Journal for Cleaner Production*, 17 (16): 1385–92.

Chafe, Z. (2005) "Consumer Demand and Operator Support for Socially and Environmentally Responsible Tourism," Center of Ecotourism and Environmentally Responsible Tourism. Online, available at: www.travelersphilanthropy.org/resources/documents/consumer_demand_report.pdf (accessed September 24, 2012).

Consultancy and Research for Environmental Management (CREM)(2000) "Feasibility and Market Study for a European Eco-label for Tourist Attractions," Consultancy and Research for Environmental Management. Online, available at: www.ecosmes.net/cm/retreiveATT?idAtt=3391 (accessed November 10, 2012).

Department for Transport (2007) "Public Attitudes Towards Climate Change and the Impact of Transport," UK Department for Transport. Online, available at: www.gov.uk/government/publications/public-attitudes-to-climate-change-and-the-impact-of-transport-in-2011 (accessed December 12, 2012).

Dickinson, J.E. and Dickinson, J.A. (2006) "Local Transport and Social Representations: Challenging the Assumptions for Sustainable Tourism" in *Journal of Sustainable Tourism*, 14 (2): 192–208.

Dickinson, J.E. and Robbins, D. (2008) "Representations of Tourism Transport Problems in a Rural Destination" in *Tourism Management*, 29 (6): 1110–21.

Eligh, J., Welford, R. and Ytterhus, B. (2002) "The Production of Sustainable Tourism: Concepts and Examples from Norway" in *Sustainable Development*, 10 (4): 223–34.

European Commission (1998) "Facts and Figures on the Europeans on Holiday," European Commission. Online, available at: www.ec.europa.eu/public_opinion/archives/ebs/ebs_117_en.pdf (accessed August 22, 2012).

Fairweather, J.R., Maslin, C. and Simmons, D.G. (2005) "Environmental Values and Response to Eco-labels Among International Visitors to New Zealand" in *Journal of Sustainable Tourism*, 13 (1): 82–98.

Ferhan, G. (2006) "Components of Sustainability: Two Cases from Turkey" in *Annals of Tourism Research*, 33 (2): 442–55.

Forsyth, T. (1997) "Environmental Responsibility and Business Regulation: The Case of Sustainable Tourism" in *Geographical Journal*, 163 (3): 270–80.

Garvill, J., Marell, A. and Nordlund, A. (2003) "Effects of Increased Awareness on Choice of Travel Mode" in *Transportation*, 30 (1): 63–79.

Gössling, S. (2009) "Carbon Neutral Destinations: A Conceptual Analysis" in *Journal of Sustainable Tourism*, 17 (1): 17–37.

Götz, K., Loose, W., Schmied, M. and Schubert, S. (2002) "Mobility Styles in Leisure Time: Reducing the Environmental Impacts of Leisure and Tourism Travel" in *Öko-Institut e.V.* Online, available at: www.strc.ch/conferences/2008/2008_OhnmachtMobilityStyles.pdf (accessed February 11, 2013).

Gronau, W. and Kagermeier, A. (2007) "Key Factors for Successful Leisure and Tourism Public Transport Provision" in *Journal of Transport Geography*, 15 (2): 127–35.

Grossberg, R., Treves, A. and Naughton-Treves, L. (2003) "The Incidental Ecotourist: Measuring Visitor Impacts on Endangered Howler Monkeys at a Belizean Archaeological Site" in *Environmental Conservation*, 30 (1): 40–51.

Hares, A., Dickinson, J. and Wilkes, K. (2010) "Climate Change and the Air Travel Decisions of UK Tourists" in *Journal of Transport Geography*, 18 (3): 466–73.

Hjalager, A. (1999) "Consumerism and Sustainable Tourism" in *Journal of Travel & Tourism* Marketing, 8 (3): 1–20.

Hudson, S. and Ritchie, J.R.B. (2001) "Cross-Cultural Tourist Behaviour: An Analysis of Tourist Attitudes Towards the Environment" in *Journal of Travel & Tourism Marketing*, 10 (2): 1–22.

Hudson, S. and Miller, G.A. (2005) "The Responsible Marketing of Tourism: The Case of Canadian Mountain Holidays" in *Tourism Management*, 26 (2): 133–42.

Huesemann, M.H. (2002) "The Inherent Biases in Environmental Research and their Effects on Public Policy" in *Futures*, 34 (7): 621–33.

Jeng, J. and Fesenmaier, D.R. (2002) "Conceptualizing the Travel Decision-Making Hierarchy: A Review of Recent Developments" in *Tourism* Analysis, 7 (1): 15–32.

Jensen, S., Birch, M. and Fredriksen, M. (2004) "Are Tourists Aware of Tourism Eco-labels? Results From a Study in the County of Storström in Denmark," 13th Nordic Symposium in Tourism and Hospitality Research, December 9, 2004.

Kaae, B.C. (2001) "The Perceptions of Tourists and Residents of Sustainable Tourism Principles and Environmental Initiatives" in S.F. McCool and R.N. Moisey (eds.), *Tourism, Recreation and Sustainability: Linking Culture and the Environment*, Wallingford, Oxfordshire: CAB International: 289–314.

Kahneman, D. (2003) "Maps of Bounded Rationality: Psychology for Behavioural Economics" in *The American Economic Review*, 93 (5): 1449.

Kahneman, D. and Thaler, R.H. (2006) "Anomalies: Utility Maximization and Experienced Utility" in *The Journal of Economic Perspectives*, 20 (1): 221–34.

Kang, M. and Moscardo, G. (2006) "Exploring Cross-cultural Differences in Attitudes Towards Responsible Tourist Behaviour: A Comparison of Korean, British and Australian Tourists" in *Asia Pacific Journal of Tourism Research*, 11 (4): 303.

Khan, M. (2003) "ECOSERV: Ecotourists' Quality Expectations" in *Annals of Tourism Research*, 30 (1): 109–24.

McKercher, B., Prideaux, B., Cheung, C. and Law, R. (2010) "Achieving Voluntary Reductions in the Carbon Footprint of Tourism and Climate Change" in *Journal of Sustainable Tourism*, 18 (3): 297.

Mihalic, T. (2001) "Environmental Behaviour Implications for Tourist Destinations and Eco-labels" in X. Font and R. Buckley (eds.), *Tourism Ecolabelling: Certification and Promotion of Sustainable Management*, Wallingford, Oxfordshire: CAB International: 57–70.

Miller, G. (2001) "The Development of Indicators for Sustainable Tourism: Results of a Delphi Survey of Tourism Researchers" in *Tourism Management*, 22 (4): 351–62.

Mustonen, P. (2003) "Environment as a Criterion for Choosing a Holiday Destination: Arguments and Findings" in *Tourism Recreation Research*, 28 (1): 35–46.

Peattie, K. (2001) "Golden Goose or Wild Goose? The Hunt for the Green Consumer" in *Business Strategy and the Environment*, 10 (4): 187–99.

Pizam, A. and Mansfeld, Y. (eds.) (1999) *Consumer Behaviour in Travel and Tourism*, New York: Haworth Hospitality Press.

Poon, A. (1994) "The 'New Tourism' Revolution" in *Tourism Management*, 15 (2): 91–2.

Reiser, A. and Simmons, D.G. (2005) "A Quasi-experimental Method for Testing the Effectiveness of Eco-label Promotion" in *Journal of Sustainable Tourism*, 13 (6): 590–616.

Saarinen, J. (2006) "Traditions of Sustainability in Tourism Studies" in *Annals of Tourism Research*, 33 (4): 1121.

Salo, M., Lähteenoja, S. and Lettenmeier, M., (2008) "Natural Resource Consumption of Tourism: Case Study on Free Time Residences and Hotel Accommodation in Finland," presented at 2nd Conference of the Sustainable Consumption and Production Research Exchange (SCORE!) Network, Brussels, Belgium, March 11–12.

Schianetz, K., Jones, T., Kavanagh, L., Walker, P.A., Lockington, D. and Wood, D. (2009) "The Practicalities of a Learning Tourism Destination: A Case Study of the Ningaloo Coast" in *International Journal of Tourism Research*, 11 (6): 567–81.

Sharpley, R. (2001) "The Consumer Behaviour Context of Ecolabelling" in X. Font and R. Buckley (eds.), *Tourism Ecolabelling: Certification and Promotion of Sustainable Management*, Wallingford, Oxfordshire: CAB International: 41–56.

Swarbrooke, J. and Horner, S. (2001) *Consumer Behaviour in Tourism*, 2nd edition, Oxford: Butterworth-Heinemann.

TripAdvisor (2011) "Travellers Keen on Going Green." Online, available at: www.tripadvisor.com/PressCenter-i134-c1-Press_Releases.html (accessed March 15, 2012).

UNEP and UNWTO (2008) "Climate Change and Tourism – Responding to Global Challenges," UNEP. Online, available at: www.unwto.org/sdt/news/en/pdf/climate2008.pdf (accessed October 7, 2012).

UNWTO (2004) "Indicators of Sustainable Development for Tourism Destinations: A Guidebook," UNWTO. Online, available at: www.uneptie.org/PC/tourism/library/goodpractices-hotel.htm (accessed November 14, 2012).

Wurzinger, S. (2003) "Are Ecotourists Really More 'Eco'? A Comparison of General Environmental Beliefs, Specific Attitudes, General Ecological Behaviour, and Knowledge Between Swedish Ecotourists and Non-Ecotourists," Lund University, School of Architecture.

Wurzinger, S. and Johansson, M. (2006) "Environmental Concern and Knowledge of Ecotourism Among Three Groups of Swedish Tourists" in *Journal of Travel Research*, 45 (2): 217–26.

Zilinger, M. (2005) "A Spatial Approach to Tourists' Travel Routes in Sweden," European Tourism Research Institute. Online, available at: www.diva-portal.org/smash/get/diva2:1380/FULLTEXT01 (accessed February 7, 2013).

7 Tourism's relationship with ethical food systems
Fertile ground for research

Carol Kline, Whitney Knollenberg and Cynthia Shirley Deale

Introduction

In society in general, and in the tourism industry, there is increased focus on local and sustainable foods. One does not have to look far to observe demonstrations of this trend, such as farmers markets, farm-to-fork restaurants, the consideration of food miles, labeling that addresses sustainability, and movements such as Slow Food, Fair Trade, and various "eat local" campaigns (Hall and Gössling 2013). As these food-related activities become more evident, they are also seen in tourism, which centers heavily on food, thus creating "new foodways and commodity chains" as noted by Hall and Gössling (2013) in their newly edited volume entitled *Sustainable Culinary Systems: Local Foods, Innovation, Tourism and Hospitality*. This chapter examines the ethical challenges present at each stage of the food system and their connections between tourism and sustainable food systems. It also illuminates various stakeholder perspectives and identifies research questions related to the connection in tourism between sustainable food and ethical consumption.

In this context, the sustainable food system is defined as "a collaborative effort to build more locally based, self reliant food economies – one in which sustainable food production, processing, distribution, and consumption is integrated to enhance the economic, environmental and social health of a particular place" (Feenstra 2002: 100). Sustainable tourism in its broadest sense is "tourism which meets the needs of present tourists and host regions while protecting and enhancing opportunity for the future" (World Tourism Organization 1993: 7), while specifically placing emphasis on people, their communities, customs and lifestyles, and ensuring they share in the economic benefits of tourism (Eber 1992). Sustainable actions within both systems are defined and discussed throughout this chapter. While ethical and sustainable behaviors are not synonymous, the adoption of sustainable actions can foster an ethical food system, and increase ethical consumption within a tourism experience.

While research in both tourism and agriculture has begun to focus on sustainability issues, the realms of sustainable and ethical activity within food systems and tourism are not yet working in tandem, despite their interdependence and obvious philosophical agreement. To address this gap, the authors outline areas where tourism stakeholders might interface with the food system (see Figure 7.1). Tourism experiences occur at each stage of the food supply chain, thereby providing opportunities for identifying ways to improve the sustainability and ethical nature of the food tourism experience.

This chapter intermeshes stakeholder theory with the basic stages of supply chain theory in the context of food: growth/production, harvesting, processing, adding value, packaging, distribution and consumption (Pullman and Wu 2011). A variety of stakeholders, defined by Freeman (1984) as any group or individual which can affect or is affected by an organization, are present at each stage and may include: visitors, residents, hosts, producers, distributors and processors. Throughout the sustainable food system these stakeholders face ethical decisions related to food and can influence tourists' access to, and ethical decisions about, sustainable

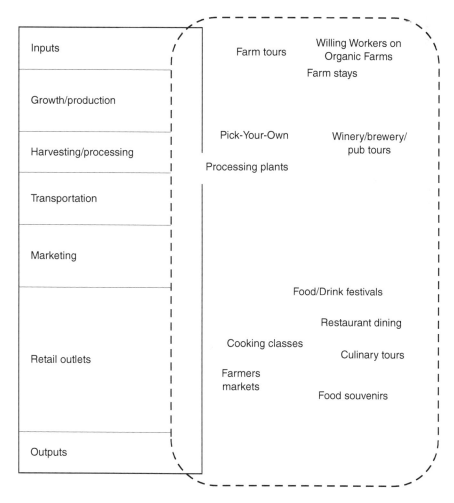

Figure 7.1 Interface of tourism experiences and the food supply chain.

food. The combination of these two theoretical perspectives provides a unique lens through which to examine the ethical implications of food tourism experiences, and introduces research questions focused on this increasingly important area.

Growth/production

Ethical decisions about food consumption begin at the growth and production stage. Books such as Pollan's *The Omnivore's Dilemma* (2006) and Schlosser's *Fast Food Nation* (2001) have raised public awareness about the social, environmental, economic and political damage that agriculture can have on individuals and communities, encouraging a growing number of consumers to consider sustainable farming practices in their decisions related to food consumption at home and while traveling (Smith 2008). Modern society's persistent demand for out-of-season and "cheap" food has supported a shift to industrial scale growth and production, which can have a detrimental impact on resources whose quality is important to both tourism and local residents. Environmental resources such as water, soil and air are degraded by the use of pesticides (Kellogg *et al.* 2000) and fertilizers (Wlederholt and Johnson 2005) in crop production, and by the amount of animal waste created when livestock are raised in Confined Animal Feeding Operations (Environmental Protection Agency 2011). Industrial-scale farming may also have negative economic and social impacts on a community (Lobao and Meyer 2001), such as the payment of low wages, hiring a high number of undocumented immigrants, and the creation of unsafe working conditions for employees, all of which has resulted in more toxic chemical injuries to workers being recorded than in any other sector of the US economy (Liu and Appollon 2010).

Tolbert *et al.* (1998) found evidence that the presence of family farms provides socioeconomic benefits to communities which may encourage citizens to stay in the community, whereas the shift to industrialized food growth and production has led to the loss of family farming traditions. Tourism focusing on the growth and production of food in a sustainable manner may provide opportunities for farmers to showcase traditional agricultural knowledge, educate consumers about sustainable agriculture, offer new avenues for revenue generation, and encourage younger generations to continue to participate in family farming.

The way food is produced can respect environmental resources, protect the people involved in its production, and provide economic support to farming communities. The decisions made by consumers about their food at home, or while travelling, have ethical implications. It is therefore important to understand how and why they are making those choices. Future research questions related to the growth and production stage of the food tourism system include:

- To what degree do tourists assess the potential environmental, social and economic impacts of the growth and production process when making decisions about food consumption?
- How effective is education, through agritourism, in causing a change in purchasing habits?

Harvesting, processing, adding value and packaging

Visitors may be interested in observing or participating in the harvesting and processing of some foods. Such activities represent potential co-creation of experiences, where tourists are actively involved in the creation of their tourism experience (Binkhorst and Dekker 2009). Some interactive food tourism opportunities have been studied, such as factory tours (Boniface 2003), and tours of wineries (Barber *et al.* 2010) and breweries (Plummer *et al.* 2005). Culinary classes and culinary vacations add value and are increasing in popularity (Smith and Xiao 2008). Pick-your-own (PYO) operations vary in their environmentally sustainable practices, and in the processing stage, animal welfare is also a critical issue.

Other key issues in agricultural processing are working conditions and payment. The tourism industry has been cited as a low-wage industry with poor working conditions for those employed at the margins; so too has agriculture. Issues in the ethical food movement include low worker pay, poor working conditions and the treatment of minorities (Liu and Appollon 2010). The Fair Trade label is considered increasingly in the tourism community as a way to ameliorate some of these issues (Boluk 2011).

According to the Food Dollar Series of the USDA's Economic Research Service, 11.6 percent of every dollar associated with food is actually earned by farms and agricultural businesses (Canning 2011). Another 18.6 percent goes to food processing, 7.5 percent to packaging and transportation, 13.6 percent to the retail trade and 33.7 percent to food services. Just under 15 percent is represented by energy costs, financing, insurance and other expenses. Potentially, if the amount of processing, transportation, and packaging required to get the food to the consumer were reduced, the farmer could capture a greater profit margin. In addition, packaging and the use of water and energy during food processing have significant effects on the overall sustainability, and, therefore, the ethics of the final food product (Chi *et al.* 2010). Value-added food products that visitors purchase as a souvenir or gift are processed in a plant outside the destination area, compromising the authenticity of an experience gained through the produce and perpetuating economic leakage. While often these components are out of view of the visitor, the savvy sustainably minded consumer would consider these processing practices when judging a food experience as sustainable or ethical. Some future research questions related to harvesting, processing and packaging include:

- What current best practices exist between the food, tourism, labor and racial justice movements?
- To what extent are visitors knowledgable about the promotion of products that are not genuinely grown, produced and packaged at the tourism destination?
- How might tourism activities that allow for co-creation, such as cooking schools or pick-your-own operations, target tourists interested in ethical food systems?

Transportation and distribution

Eating local food can reduce the need for packaging, reduce carbon emissions associated with transporting food, and increase the level of food security. However, some studies note that transportation, in comparison to other energy needs in food production and processing, may not be a major culprit contributing to carbon emissions. A 2005 study for the UK Department for Environment, Food and Rural Affairs, noted that while sea freight accounted for 65 percent of UK food transport, it was responsible for only 12 percent of the carbon dioxide emissions associated with that transport. In fact, transportation accounts for just 10 percent of emissions associated with the UK's food chain (Chi *et al.* 2010). Some researchers are critical of the "food miles" approach to sustainability (Chi *et al.* 2010), noting that other factors involved in food matter more in terms of their environmental impact. Certification programs, such as those offered by the Food Alliance, have been developed as ways to work toward evaluating management practices associated with these processes. Future research questions related to transportation and distribution include:

- Would creating a greater domestic demand for local food through tourism help to reduce emissions associated with food transportation? Would this impact the availability of local products for residents?

Community destination

A community's cultural and recreational activities, or a vibrant downtown, can be enjoyed by both residents and visitors, as can its infrastructure, such as well-maintained roads and adequate medical and banking services. Both are inherent to a community's quality of life and are a measure of its basic vitality. So too is the health of a destination's residents. While evidence of an obesity epidemic in the USA is well documented in the nutrition literature, the connection between the health of a destination's population and the visitor experience is absent in the tourism literature.

A community-based issue related to distribution has been dubbed "food access," or the ability of low-income residents to find and purchase healthy food. In fact, communities without access to a full range of groceries have been called "food deserts" (USDA 2012). Food access may not appear to relate to tourism; however, in the broadest sense, it affects the community upon which the destination rests, and is concurrent with the disparities between the lifestyles of residents and visitors that is documented within tourism literature (Fagence 2003). Voluntourism may provide an opportunity for tourism to positively benefit food access in rural or urban environments. Future research questions related to a destination community's environment include:

- To what extent do tourists notice a sense of community health/vibrancy, and in what ways does it matter to them when traveling?

- How can tourism also provide an avenue to improve the availability of healthy, nutritional, affordable and sustainably raised food for residents?
- How successfully are tourism development groups able to tap into food-related funding for support of projects?
- What are the current best practices for sustainable farm, food, and food access groups to work with tourism representatives to enhance the food system?
- How can the overlap between voluntourism experiences and the sustainable food system be developed to amplify ethical consumption in tourism?

Marketing

Research on food tourism marketing has generally focused on branding (Boyne and Hall 2004), accurate representations of food destinations (Frochot 2003), and the benefits of promoting local foods to residents and tourists (Carlsen and Edwards 2008). While several of these studies concluded that promoting unique food and food production methods allows destinations to distinguish themselves from others, Frochot (2003) found that some destinations do not capture the full marketing potential of their unique food systems.

Others may misrepresent their food systems, which as Boyne and Hall (2004: 82) suggest, may lead to the creation of an unsustainable food tourism destination. False promotion of sustainable practices has led to consumer skepticism. Efforts have been made to regulate claims of sustainability through the use of labels or certifications such as "US Department of Agriculture (USDA) Organic" or "Fair Trade." But labels may vary between jurisdictions, therefore tourists who use these labels and certifications to aid their ethical decision-making should understand their requirements. Although researchers have investigated the connection between the marketing stage of the food system and tourism, more work could be done in the following areas:

- To what degree do tourists rely on labeling and certifications to make ethical decisions about the destinations they visit and the food they consume while traveling?
- What types of marketing efforts best boost the visibility of sustainable products?

Consumption/consumer choice while traveling

In addition to nutrition, taste, freshness and variety, an ethical eater may consider a food's level of local-ness, seasonality, pesticide use, Fair Trade, animal welfare and/or assistance to a developing nation. In particular, the current rate of meat consumption and its methods of production are not considered to be sustainable (Hoogland *et al.* 2005). Chi *et al.* (2010: 43) suggest "Meat and dairy portions is perhaps the most significant action you can take to reduce the impact of food production on people and planet."

While the demand for locally grown food has increased, sustainable food choices while traveling is an area ripe for exploration. In a study on product selection, Batte *et al.* (2010) found that in addition to price, recognition of a national brand, support for small family farms, organic certification labeling and text, and regional and state product identification all significantly increased the selection of a product and/or the willingness to pay more for that product.

Profile of food tourists

Mitchell and Hall (2003) noted that food tourists were well educated, attracted to new and innovative things, and sought knowledge and education. However, they only briefly mentioned food tourists who were "socially aware," noting that they were more likely to be motivated by wine and food tourism products. According to the same authors, food tourists seek different dining experiences from their residential counterparts and possess different consumption patterns while traveling than when they are at home. Additionally, they suggest a typology of food tourist behavior along a continuum from neophilia (love of the new) to neophobia (fear of the new) where *Gastronome* are the most adventurous with high interest and high involvement in food experiences while traveling, engaging in cooking schools, food education, "high cuisine," farmers' markets, and local growers and suppliers. Alternatively, the *Familiar Foods* category represents travelers with low interest, involvement and risk in their away-from-home food experiences. A variety of other issues may influence consumer choice, including religion (Wan Hassan and Hall 2003) and environmental accounting in food choice (Chi *et al.* 2010). Future research questions related to consumer choice include:

- At what planning stages do tourists make their decisions about engaging in food tourism experiences? Which stages have the strongest influences on ethical consumer behavior?
- How much do experiences while traveling affect a visitor's consumption patterns after the trip is completed?
- While the connection between religion and travel has been explored in studies regarding pilgrimages, religious heritage sites and traveler conflicts, how do visitors who eat according to a religious doctrine approach food tourism experiences?

Consumable product outlets

The majority of literature on food tourism focuses on consumable product outlets (CPOs); places where visitors (and residents) purchase and consume food products. While some direct consumption occurs on farms, sustainable food experiences generally take place in other locales including (on- or off-farm) produce stands, farmers markets, food trucks, farm-to-fork restaurants, farm-to-fork dinners on farms, and grocery or specialty retail stores. At many of these locales, raw, prepared, and/or value-added products can be purchased and consumed.

Raw products would include produce, herbs, eggs and dairy items, and sustainably raised meat. Prepared products might include restaurant meals or food samples or tastings of cooked items. Value-added products reflect foods that are also prepared, but have a shelf-life, such as jams, jellies, cheeses, salsas and cured meats. Numerous research questions remain about CPOs within an ethical tourism experience context; several will be identified in the following paragraphs.

Restaurants

The value added to food in a restaurant setting has been researched in the tourism literature through niches such as culinary tourism, wine tourism, beer tourism, food heritage, artisanal and regional foods, and slow foods. Restaurants serve as the most conspicuous place for tourists to make ethical consumption decisions because they are frequently the most accessible option. Yet, as more restaurants establish themselves as members of the sustainable food system, it is important that tourists assess the information provided by the restaurant. Some restaurateurs list farms and production facilities that source their ingredients and this is a crucial part of the dining experience, while others choose not to tout their suppliers. Should diners continue to request this information, and if the demand for sustainably sourced and produced food continues to grow, more pressure may be placed on restaurateurs to provide sustainable and ethical options for their patrons. As a ubiquitous component of the food system, restaurants represent an important area for further research and questions include:

- How and when do consumers gather information about their ethical consumption options in restaurants?
- To what extent does consumer demand influence a restaurateur's/chef's decisions to use sustainably sourced or produced products?
- How does offering sustainably sourced and produced food such as Fair Trade, organic, and local items influence the profit of a restaurant?

Meetings/conventions/events/caterers

Meeting Professionals International, a major meetings professional organization, promotes sustainability through offering sustainable foods at events, and has launched a sustainable meeting measurement toolkit. Green Planet Catering engages in sustainable practices related to food by sourcing it from local farmers, delivering food in biodiesel-powered trucks, serving on biodegradable plates, and picking up the waste from an event to make sure it is composted. Research questions regarding meetings, conventions, special events and catering include:

- What sustainable foods, and in what amounts, are currently offered at meetings? Which sustainable items could be promoted more heavily, and what gaps exist between products that could be offered?

- How can meeting planners work better with caterers to ensure that sustainable foods are offered?

Retail outlets

Gift shops, grocery stores, produce stands and farmers markets are venues for visitors to purchase value-added agricultural products reflecting the region visited. Ethical issues concerning retail outlets may include: type of establishment (i.e. convenience store retail outlets typically focus on convenience and not on ethical consumption or nutrition), ownership and management of establishment, treatment of employees, training of employees, and percentage of locally made products. A research question that might be asked is the following:

- What is the comparative demand for sustainable food by residents and visitors in varied retail outlets?

Food trucks

According to the US National Restaurant Association Consumer Survey (August 2011), 18 percent of Americans saw a food truck in their community during the summer of 2011, and 28 percent of those who saw one made a purchase from it. As a relatively new supplier, the following questions can be asked about food trucks:

- What kinds of sustainable products (e.g. organic, local, pasture-raised meats, certifications) are typically offered by food trucks?
- To what extent do food truck patrons apply their sense of ethics or values when making purchasing decisions at a food truck?
- Do tourists perceive or seek out food trucks as a source of food that supports their ethics and values?

Festivals

Festivals have been studied as a means of supporting a sustainable economy (O'Sullivan and Jackson 2002), preserving heritage (Chhabra *et al.* 2003), and as important activities for farmers (Barbieri and Mshenga 2008), supporting McGehee and Kim's (2004) finding that a majority of agritourism operators engage in on-site festivals. However, little work has examined relationships between sustainable food systems and festivals. From the participant's perspective, Chang and Yuan (2011) found that food tasting, trying region-specific foods, and increasing knowledge of food were important variables in tourists' motivations to attend food festivals. To further understand the interactions between sustainable food producers and festival attendees, research questions include:

Tourism and ethical food systems 113

- To what extent do festival attendees (both resident and visitor) seek out sustainable food producers at festivals?
- Can interactions with sustainable food producers at festivals change consumers' attitudes or behaviors related to sustainable food?

Food or agricultural tours

Sometimes CPOs are packaged together into guided culinary tours. Regional winery tours are examples that often include products with specific sustainability designations such as LIVE (Low Input Viticulture and Enology). A unique example of a tour is offered by Tastes of Portland; the company partners with local food businesses to offer cyclists a taste of food from several vendors. Research is needed on this growing segment of the tourism industry. Questions include:

- Can experiences on farm or food tours change consumers' attitudes or behaviors related to ethical food decisions?
- Do sustainable food or agricultural tours increase brand recognition and profits for these products/farms?

Outputs/waste

Waste and by-products, including chemical run-off, various forms of pollution, water waste, packaging waste and food waste, occur along each stage of the food supply chain; minimizing waste and its associated cost makes sense for the economy and for the environment. Uneaten food in the US is valued at $165 billion, and food waste is the single largest component of solid waste in US landfills (Gunders 2012), while more than 3.5 million tons of discarded food is collected annually by local waste authorities in the UK (Chi *et al.* 2010). However, food waste has not been a main focus of businesses or government until recently. In 2012, a food waste reduction conference was held in North Carolina, USA. (Carolina Recycling Association n.d.), targeted at business and government leaders, restaurateurs, hospital administrators, hauliers, composters, and non-profit agencies who want to create successful food waste diversion programs. As additional events like this one evolve, several questions come to mind. These include:

- How many (if any) tourism businesses and related agencies will become involved?
- What potential role can tourism play in diminishing the destructive waste of industrial food systems, or is it actually an ethical issue at all?

Discussion and research directions for the future

At each stage of the food value chain, stakeholders contribute to the continuation, advancement, or expansion of the sustainable food system, and without stakeholder cooperation and communication the system may fail (Andereck and Vogt 2000). Tourists engage in decisions about where to eat, what to purchase and where to visit. Upon returning home they may decide to seek out food products from the destination they visited. These stakeholders control the demand for sustainable food products and experiences in tourism systems. It is critical to ensure that they are aware that they have the power to make ethical consumption decisions. Destination residents, like visitors, are consumers of food products and should also be consulted in decisions related to how ethical food products should be made accessible to consumers. Hosts are involved in destination marketing organizations, such as tourism planners, and in a variety of tourism-related businesses. Hosts face an ethical dilemma related to the creation of a destination's image based on food associated with that destination; this image must be accurate or the efforts will be unsustainable. Finally, farmers, wholesale and retail distributors, and processors such as chefs and food artisans, create the supply side of the sustainable food system. These stakeholders help drive the sustainable food system and can aid or hinder the tourist's ability to make ethical decisions about the foods they consume.

Although the literature provides a foundation for understanding connections between food and tourism systems and food and tourism stakeholders, more research is needed to fully understand the synergy between them. Attention to the supply chain allows researchers to carefully examine each component of the food value chain, for every stage contributes to the sustainability of the final product. A focus on stakeholders then mandates consideration of every party impacted, and consequently provides a holistic understanding of the two systems as well as the connections between them. These theories have proven useful when using a wide lens to examine the systems, but more specific theories such as Flora's community capitals framework (2012), social capital theory, Ateljevic's circuits of production/consumption (2000), or Theory of Reasoned Action (Fishbein and Ajzen 1975) would all lend valuable perspectives and insights toward understanding the development of ethical tourism experiences along the food supply chain. Entrepreneurship plays an important role in the creation of innovative food programs and/or tourism products, and therefore theories such as Sundbo et al.'s notions of co-creation (2010), Santos's theory of social entrepreneurship (Santos 2009), and the ecological systems theory (Bronfenbrenner 1979) which takes into account the natural, political, social, financial and human systems that interact to bring about change, may prove useful in exploring the research questions identified in this chapter.

As illustrated in this chapter, each stage of the food system offers opportunities for tourism to help improve its sustainability. The tourism industry has not yet made significant efforts to support sustainable food systems around the globe; many so-called "green" hoteliers continue to source food from industrial food systems, while providers of local food experiences fail to incorporate "green" or Fair Trade practices. The tourism industry must adopt a more comprehensive approach to sustainability to integrate food with multiple aspects of sustainability. The sustainable and local food movement, which is linked with tourism, must

Tourism and ethical food systems 115

partner with tourism entities to develop, accept and practice a code of ethics with regard to energy use, working conditions, packaging, food access, marketing and waste. Many overarching questions remain about ethical food tourism experiences; therefore, several directions for research are presented below.

Food purchase and consumption behavior while traveling

Basic knowledge about food purchase and consumption behavior away from home is needed and can be collected through answering the following:

- Do tourists have different food purchase and consumption behaviors when they travel than they do at home? If so, how do they differ, and why?
- Do tourist markets shift along the food system? In other words, are the tourist markets that visit farms similar to the markets participating in culinary classes or visiting wineries?

Ethical decision-making regarding food

Visitors who travel with ethical considerations in mind are presented a dizzying array of choice regarding the economic, social and environmental impacts of their trip experiences. Questions for research in this area include:

- How do visitors make decisions about the food they eat while traveling? In what ways do ethical food considerations compare with other consumption decisions?
- To what extent and in what ways are visitors influenced by educational or marketing messages related to ethics, sustainability, community and other characteristics communicated about the food system?

Tourism's contribution to sustainable and ethical food systems

An overarching theme in this chapter is speculation over tourism's ability to enact change in a destination community. The following research questions promote future inquiry into this issue:

- What are the greatest challenges facing sustainable food production, and how might an emphasis on ethical tourism consumption play a role in breaking down some of the barriers?
- How might the tourism industry assist small food businesses in scaling up, a classic production issue in small to moderate businesses?
- How does an increase in food tourism impact residents' quality of life?
- Does the tourism industry, and more specifically do the stakeholders within it, have an ethical responsibility to improve the food supply chain? How can tourism organizations partner to develop effective regional, sustainable, food tourism strategies?

- How can the tourism industry best develop and disseminate clear messages about sustainable food products, experiences and ordering processes?
- What are new ways that tourism providers can foster the "packaging" of existing market-ready and near market-ready sustainable food products and experiences?
- What are examples of where the tourism industry has partnered with the sustainable food industry to initiate pilot projects to develop new sustainable food products and visitor experiences? What were the antecedents to the initiation of these projects?

Destination community environment

Questions regarding where the destination ends and the community begins have been treated in tourism literature for some time. This focus can inspire productive inquiry, bridging policies on public health, food access, entrepreneurship, civic involvement and regional infrastructure development. Questions guiding such future studies include:

- Do visitors notice or care about the physical health of a destination's population? If they notice, how do these impressions enhance or detract from their perception of the destination or of their own travel experience?
- Which policies that support a strong entrepreneurial community also enhance food access and public health?
- What are the most critical infrastructure and policies needed in a destination community by stakeholders to make ethical decisions regarding food growth, production, distribution and consumption?

Stakeholders within food and tourism

A greater understanding of the perspectives of each stakeholder group, and the interactions between them, would inform planners and managers. Questions guiding such inquiry include:

- How do stakeholder groups within the food tourism supply chain view their own ethical landscape? How do they view the ethical landscape of others at different stages in the supply chain?
- What roles do social networks and social capital play in expanding the linkages between and innovation within sustainable food systems and sustainable tourism?

Theoretical contributions to ethical consumption

Finally, for a global understanding of ethical food tourism consumption to grow, the theoretical foundation of current research needs to be analyzed, leading to questions such as the following:

- Which theoretical frameworks have dominated the food tourism literature to date?
- Which particular theories lend themselves better to each stage of the food tourist experience than others? Do particular theories resonate best with ethical decision-making of tourists? Of other stakeholders?

Conclusion

This chapter combined supply chain theory and stakeholder theory to develop a multi-dimensional examination of ethical food and tourism. Ultimately, the goal of sustainable food tourism can be supported by organizations such as the Sustainable Table, an organization that communicates about sustainable eating, and summed up by the London-based organization Sustain that "advocates food and agricultural policies and practices that enhance the health and welfare of people and animals, improve the working and living environment, promote equity and enrich society and culture." After all, ethical food is an issue for all stakeholders, and tourism has the opportunity to be a conduit and an innovator in terms of increasing awareness, improving education and influencing behavior toward an ethical and sustainable future for food globally.

Activity

Tourism and your local food system

Investigate sustainable food in your own community and try to determine who produces it and when, where, and how is it available to you and to visitors?

Contemplate the definition and stages of the food system, and determine ethical food consumption decisions that visitors to your community could encounter. Consider the following:

- What farms or markets could they visit and where else would these products be available in your community? How far would visitors have to travel for these products?
- Are there restaurants, hotels, festivals and events, or other venues where visitors could find sustainable food products?
- What sort of marketing material, or other information, is available to help visitors make ethical decisions?
- Are there value-added products for visitors to purchase as souvenirs; if so, what are those products and where would they find them?
- What sorts of measures is your community taking to reduce food waste?

Now consider sustainable food system stakeholders: visitors, residents, hosts, producers, processors and distributors. Review this chapter's definition of sustainable tourism, and determine what their roles in sustainable tourism development might be. Then, brainstorm specific examples of how visitors to your community could influence them by making ethical decisions regarding their role in food production and consumption. Are there others who may be directly (or indirectly) influenced by such decisions? How would each set of stakeholders be influenced?

To conclude, think about how easy or difficult it would be for a visitor to make ethical decisions about food consumption in your community. Could barriers such as access, distance, policies or price keep visitors and residents from making ethical decisions about their food consumption? Could anything be done to make it easier for visitors (and residents) to be more ethical in their food consumption? Who would stand to benefit or lose from those changes?

Recommended reading

To gain a deeper understanding of how sustainable tourism can provide ethical working conditions for farm workers and others in the food system please see:

Liu, Y.Y. and Apollon, D. (2011) *The Color of Food*, New York: Applied Research Center. Online, available at: http://arc.org/downloads/food_justice_021611_F.pdf.

A Short Guide To Sustainable Agriculture outlines how sustainable agriculture can be used to address some of the ethical issues associated with industrial-scale farming:

Halliday, J. and Schuttelaar & Partners (June 2009) *A Short Guide To Sustainable Agriculture*. Hartland, VT: SAI Platform and Sustainable Food Lab. Online, available at: www.sustainablefood.org/images/stories/pdf/SFL-SAI%20Sustainable%20Ag%20guide%20final.pdf.

The following case study provides an example of how domestic markets (including tourism providers) can help support the sustainable food system:

Mariano, A., Uy, J., Mendoza, L., Tuason, P.T., Cuyco, O., Torres, V. and Tiongson, G. (2011) *Farmer Entrepreneurship Program: Experience of small farmer clusters in Nueva Ecija, Philippines*. Online, available at: www.linkingworlds.org/images/stories/Jollibee%20Philippines_%20CRS%20case%20study-6.pdf.

References

Andereck, K.L. and Vogt, C.A. (2000) "The Relationship Between Residents' Attitudes Toward Tourism And Tourism Development Options" in *Journal of Travel Research*, 39 (1): 27–36.
Ateljevic, I. (2000) "Circuits Of Tourism: Stepping Beyond The 'Production/Consumption' Dichotomy" in *Tourism Geographies*, 2 (4): 369–88.
Barber, N., Taylor. D.C. and Deale C.S. (2010) "Wine Tourism, Environmental Concerns, And Purchase Intention" in *Journal of Travel & Tourism Marketing*, 27 (2): 146–65.
Barbieri, C. and Mshenga, M. (2008) "The Role Of The Firm And Owner Characteristics On The Performance Of Agritourism Farms" in *Sociologia Ruralis*, 48 (2): 166–83.

Batte, M. T., Hu, W., Woods, T. and Ernst, S. (2010) "Do Local Production, Organic Certification, Nutritional Claims, And Product Branding Pay In Consumer Food Choices?," Paper presented at the 2010 Agricultural & Applied Economics Association Annual Meeting, Denver, Colorado, July 25–27, 2010.

Binkhorst, E. and Dekker, T.D. (2009) "Agenda For Co-Creation Tourism Experience Research" in *Journal of Hospitality Marketing and Management*, 18 (2–3): 311–27.

Boluk, K. (2011) "In Consideration Of A *New* Approach To Tourism: A Critical Review Of Fair Trade Tourism" in *The Journal of Tourism and Peace Research*, 2 (1): 27–37.

Boniface, P. (2003) *Tasting Tourism: Travelling For Food And Drink*, Aldershot: Ashgate Publishing Ltd.

Boyne, S. and Hall, D. (2004) "Place Promotion Through Food And Tourism: Rural Branding And The Role Of Websites" in *Place Branding*, 1 (1): 80–92.

Bronfenbrenner, U. (1979) *Ecology of Human Development: Experiments By Nature And Design*. Boston, MA: Harvard University Press.

Canning, P. (2011) "A Revised And Expanded Food Dollar Series: A Better Understanding Of Our Food Costs," ERR-114, US Department of Agriculture Economic Research Service, February.

Carlsen, J. and Edwards, D. (2008) "BEST EN Case Studies: Innovation For Sustainable Tourism" in *Tourism and Hospitality Research*, 8 (1): 44–55.

Carolina Recycling Association, the NC Composting Council, US EPA Region 4, South Carolina DHEC, and NC DENR. Online, available at: www.cra-recycle.org/food-wasteconference/ (accessed December 12, 2012).

Chang, W. and Yuan, J. (2011) "A Taste Of Tourism: Visitors' Motivations To Attend A Food Festival" in *Event Management*, 15 (1): 13–23.

Chi, K.R., MacGregor, J. and King, R. (2010) *Big Ideas In Development: Fair Miles: Recharting The Food Miles Map*. London: International Institute for Environment and Development/Oxfam GB.

Chhabra, D., Healy, R. and Sills, E. (2003) "Staged Authenticity And Heritage Tourism" in *Annals of Tourism Research*, 30 (3): 702–19.

Eber, S. (ed.) (1992) *Beyond The Green Horizon: A Discussion Paper On Principles For Sustainable Tourism*. Godalming, UK: Worldwide Fund for Nature.

Ellen, S., Wiener, J.L. and Cobb-Walgren, C. (1991) "The Role Of Perceived Consumer Effectiveness In Motivating Environmentally Conscious Behaviors" in *Journal of Public Policy & Marketing*, 10 (2): 102–17.

Environmental Protection Agency (2011) *How Do CAFOs Impact the Environment?* Online, available at: www.epa.gov/region7/water/cafo/cafo_impact_environment.htm (accessed November 12, 2012).

Fagence, M. (2003). "Tourism And Local Society And Culture" in S. Singh, D.J Timothy and R.K. Dowling (eds.), *Tourism in Destination Communities*, Wallingford: Oxford University Press: 55–78.

Feenstra, G. (2002) "Creating Space For Sustainable Food Systems: Lessons From The Field" in *Agriculture and Human Values*, 19 (2): 99–106.

Fishbein, M. and Ajzen, I. (1975) *Belief, Attitude, Intention, and Behavior: An introduction to Theory and Research*, Reading, MA: Addison-Wesley.

Flora, C.B. and Flora, J.L. (2012) *Rural Communities: Legacy and Change* (4th edition), Boulder, Colorado: Westview Press.

Freeman, R.E. (1984) *Strategic Management: A Stakeholder Approach*, Boston: Pitman.

Frochot, I. (2003) "An Analysis Of Regional Positioning And Its Associated Food Images In French Tourism Regional Brochures" in *Journal of Travel and Tourism Marketing*, 14 (3/4): 77–96.

Gunders, D. (2012) *Wasted: How America Is Losing Up To 40 Percent Of Its Food From Farm To Fork To Landfill*, New York: National Resources Defence Council.

Hall, C.M. and Gössling, S. (eds.) (2013) *Sustainable Culinary Systems: Local Foods, Innovation, Tourism And Hospitality*, New York: Routledge.

Hoogland, C.T., de Boer, J. and Boersema, J.J. (2005) "Transparency Of The Meat Chain In The Light Of Food Culture And History" [Research Report] in *Appetite*, 45: 15–23.

Kellogg, R.L., Nehring, R., Grube, A., Goss, D.W. and Plotkin, S. (2000) *Environmental Indicators of Pesticide Leaching and Runoff from Farm Fields*, Washington, DC: USDA Natural Resources Conservation Service.

Lobao, L. and Meyer, K. (2001) "The Great Agricultural Transition: Crisis, Change, And Social Consequences Of Twentieth Century US Farming" in *Annual Review of Sociology*, 27: 103–24.

Liu, Y.Y. and Apollon, D. (February 2011) *The Color of Food*, New York: Applied Research Center.

McGehee, N. and Kim, K. (2004) "Motivation For Agri-Tourism Entrepreneurship" in *Journal of Travel Research*, 43 (2): 161–70.

Mitchell R. and Hall, C.M. (2003) "Consuming Tourists: Food Tourism Consumer Behavior" in C.M Hall, L. Sharples, R. Mitchell, N. Macionis and B. Cambourne (eds.), *Food Tourism Around The World: Development, Management And Markets*, Oxford: Butterworth-Heinemann: 60–80.

National Restaurant Association (August 2011), "National Restaurant Association Consumer Survey." Online, available at: www.restaurant.org/ (accessed November 5, 2102).

O'Sullivan, D. and Jackson, M.J. (2002) "Festival Tourism: A Contributor To Sustainable Local Economic Development?" in *Journal of Sustainable Tourism*, 10 (4): 325–42.

Plummer, R., Telfer, D., Hashimoto, A. and Summers, R. (2005) "Beer Tourism In Canada Along The Waterloo–Wellington Ale Trail" in *Tourism Management*, 26 (3): 447–58.

Pollan, M. (2006) *The Omnivore's Dilemma*, New York: Penguin Press.

Pullman, M. and Wu, X. (2011) *Food Supply Chain Management: Economic, Social And Environmental Perspectives*, New York: Routledge.

Santos, F. (2009) "A Positive Theory Of Social Entrepreneurship," INSEAD Working Paper Series, Fontainebleau, France. Online, available at: www.insead.edu/facultyresearch/research/doc.cfm?did=41727 (accessed June 6, 2012).

Schlosser, E. (2001) *Fast Food Nation: The Dark Side Of The All-American Meal*, New York: Houghton Mifflin.

Smith, G.B. (2008) "Developing Sustainable Food Supply Chains" in *Philosophical Transactions of the Royal Society of London*, 363 (1492): 849–61.

Smith, S.L.J. and Xiao, H. (2008) "Culinary Tourism Supply Chains: A Preliminary Examination" in *Journal of Travel Research*, 46 (3): 289–99.

Sundbo. J., Sørensen, F. and Fuglsang, L. (2010) "Innovation in the Experience Sector: Research Report 10–7," Roskilde, Denmark: Centre of Service Studies, Roskilde University.

Tolbert, C.M., Lyson, T.A. and Irwin, M.D. (1998) "Local Capitalism, Civic Engagement, And Socioeconomic Well-Being" in *Social Forces*, 77 (2): 401–27.

United States Department of Agriculture Economic Research Service (2012) "Definition of a Food Desert." Online, available at: www.ers.usda.gov/data-products/food-desert-locator/documentation.aspx (accessed October 14, 2012).

Wan Hassan, M. and Hall, C.M. (2003) "The Demand For Halal Food In Muslim Travelers In New Zealand" in C.M Hall, L. Sharples, R. Mitchell, N. Macionis and B. Cambourne (eds.), *Food Tourism Around the World: Development, Management And Markets*, Oxford: Butterworth-Heinemann: 81–100.

Wlederholt, R. and Johnson, B. (2005*) Unintended Impacts of Fertilizer and Manure Mismanagement on Natural Resources*, North Dakota State University Extension Service.

World Tourism Organization (1993) *Sustainable Tourism Development: Guide for local planners*, Madrid: WTO.

8 Travelling goods
Global (self) development on sale

Maria Koleth

Introduction

The complex motivations of development tourists offer an insight into the expanding market for ethical travel. While development tourism (which involves tourists volunteering for short periods of time with development and conservation organisations in the Third World) may involve a strong, altruistic desire to aid development efforts, it also has strong incentives for individuals in a neoliberal market. As development tourism has become increasingly affordable, the motivations of young development tourists have grown from altruism and 'alternative hedonism' (Soper 2009) to encompass formative educational projects that can enhance career prospects. Often, volunteering experiences can now be recounted in job interviews and on resumés because they reflect qualities such as resilience, initiative and leadership. While the pursuit of self-development through career advancement and educational experiences may appear to compromise the overtly altruistic motivations of development tourists, it does contribute to the embodiment of a 'good' neoliberal citizen (Simpson 2005; Lyons *et al.* 2012).

The ethical transactions that create this 'good' self in development tourism reflect changes in other industries, such as those of development and social enterprise. This chapter argues that development tourism is a form of ethical transaction that enables the creation of a 'self-as-enterprise' (Foucault 2008; McNay 2009) in a changing neoliberal economy (Vrasti 2013). The formation of a 'self-as-enterprise' in tourism significantly changes the demarcations between the commodification of different tourism products, rendering the tourist themselves both a consumer and product, simultaneously. This chapter looks at how the development tourist becomes a 'self-as-enterprise', while drawing on data from qualitative interviews with Australian development tourists conducted in 2012 as well as other studies of volunteer tourists.

The growth of development tourism is contributing to wider changes in the field of global development. Rochelle Spencer, in *Development Tourism: Lessons from Cuba*, argues that development tourism represents a notable departure from the qualified technical experts usually tasked with development initiatives and is a symptom of the development industry's need to constantly diversify its 'product' (Spencer 2010: 69). Jim Butcher and Peter Smith (2010) argue in a recent article that development tourism is part of a shift in the individual's relation to political organisations and social change. Barbara Vodopivec and Rivke Jaffe (2011), in their examination of development tourism, assert that this shift in the individual's political and social commitment is a part of the wider commodification of development in the neoliberal economy. While development began with state-sanctioned projects and public works for the betterment of Third World peoples, it is now reliant on the agency of non-governmental organisations and corporate interests (ibid.). The neoliberalisation of development automatically privileges those anti-poverty measures and those versions of doing 'good' in the Third World that support capitalist enterprise (Cameron and Haanstra 2008).

Recent studies of development tourism demonstrate that altruistic motivations are being outnumbered by egoistic motivations like self-development and socialising with peers (Mustonen 2007; Gray and Campbell 2007; Sin 2009; Coghlan and Fennell 2009; Tomazos and Butler 2010). In many ways, the complexity of tourists' motivations reflects a shift in the ethical consciousness of all contemporary tourists, in that 'the tourist today is a living ethical dilemma, steering between ethically good and bad features of social exchange' (Figueroa and Waitt 2010: 262–3). Foucauldian analyses of ethical and moral selving shed more light on the broader context for this compromised form of (self) development because they shift the dilemma from that between altruism and, its other, egoism, to one of how specific practices come to be labelled as ethical (Barnett *et al.* 2005; Boluk 2011; Vrasti 2013). This ethical or moral selving refers to 'the mediated work of creating oneself as a more virtuous person through practices that acknowledge responsibilities to others' (Barnett *et al.* 2005: 30). Such a framework of analysis recognises that it is precisely the compromise between divergent motivations that is being sold to development tourists, and that 'ethical consumers often do not see morality and selfhood as competing or conflicting, but as integrated' (Varul 2009: 183). The complex subjectification of volunteer tourism reflects the primacy of the Foucauldian self-as-enterprise in a neoliberal society based on competition, differentiation and entrepreneurship (Vrasti 2013).

Methods

This chapter draws from ongoing PhD research on volunteering and development at the University of Sydney, Australia. This research focuses on a multi-sited ethnography of development tourism, including interviews in Sydney and two periods of participant observation in development tourism programmes in Cambodia and Peru. The data referred to in this chapter is drawn from 30 interviews with returned development tourists in Sydney, Australia, carried out in early 2012. The development tourists interviewed completed their programmes of volunteering between 2010 and 2012, and took part in two interviews (each) about their experiences. The names of interview participants have been changed to preserve their anonymity. The programmes in which they took part all involved a financial payment from the volunteer to the company or organisation with which they volunteered. The organisations and companies included Habitat for Humanity, Projects Abroad, Antipodeans Abroad and i-to-i Volunteers International. This data is supplemented with those of other, published, studies of volunteer tourism.

Every self has a price

The compromise between the overtly altruistic act of development and the egoism of self-development can be seen as being productive of a 'self-as-enterprise' (McNay 2009) in a neoliberal economy. Cassie, a 20-year-old student who undertook three development tourism placements with the British company i-to-i Volunteers, did not see her trip as a compromise between two opposing imperatives. Instead, her journey was taken on the understanding that it would be an enjoyable and formative experience that would enhance her skills and turn what she described as her 'naïve' 19-year-old self into something better, someone good. Cassie described this making of a good self in the kind of recognisable, branded language that is reminiscent of job interviews; phrases such as 'I'm a people person and I take a lot of initiative....[W]hen thrown in the deep end I'll find a way, find a way around it, which is kind of what happened.' Foucault's theorisation of the 'self-as-enterprise', articulated in *The Birth of Biopolitics* (Foucault 2010), envisaged an economy in which individuals, like Cassie, 'would be encouraged to see their lives and identities as a type of enterprise' (McNay 2009: 56). The key feature of this neoliberal version of the economic agent is a self that is made through being sold.

The basis for such an economy, Foucault theorised, was a radically different kind of societal space which functioned not as a juridical framework – in which a market would have its place – but rather its opposite, a space of economic freedom that would 'act as a point of attraction for the formation of a political sovereignty' (Foucault 2010: 83). The society produced from this space of economic freedom would be an 'enterprise society' (Foucault 2010: 149) regulated only by the mechanisms of competition. With competition as the guiding principle, the good society to be developed towards is one characterised by the multiplication and differentiation of the enterprise form (Foucault 2010). The increasing popularity of development tourism is a prime example of the multiplication of the enterprise form into areas of global welfare, which were not traditionally market driven (Vrasti 2013). As is typical of the enterprise form, the popularity of development tourism is driven by its differentiation from the exploitative practices of existing mass tourist enterprises.

The self-as-enterprise form reconfigures the debate over the motivations behind development tourism because in an enterprise society the compromise between altruism and egoism helps create a more competitive enterprise. Foucault (2010) argues that it would be a mistake to see the differentiation and diversification of products, and the related importance given to multiple, decentralised forms of community action, as a rejection of capitalist norms. He goes on to elucidate that it is these norms which have changed in a society that no longer evinces a passive mass consumption but rather is based on active and differentiated competition (ibid.). The difference between development tourism and environmentally destructive mass tourism is not that development tourism is an alternative to the mainstream economy but rather that it indicates a substantive shift in that economy. The development tourist is, therefore, not particularly selfish, but rather is shaped by a society in which the altruistic rejection of mass consumption is outdated (Foucault 2010). Arguably, the motives of development tourists are centred on creating a successful enterprise in an economy where everything good, including well-developed societies, is constituted by multiple and diversified enterprises. Having raised money for the children that she was volunteering with in Peru and Ecuador, Cassie was aware that her fundraising supporters would be reading her travel blog and she worked hard to portray both herself and her trip as a successful enterprise. As she described it,

> The blog was designed with the purpose of keeping everyone happy.... I tried to make it sound like, you know, what I was doing ... [with] their money was really making a difference to their lives.... [I]t was very much for people who donated money to the kids, and it was so fantastic that they did and I couldn't have done it without their donation.

Increasingly, experiences of development tourism are being sold as necessary elements of the successful 'self-as-enterprise'. Kate Simpson (2005: 454), in her analysis of the professionalisation of the volunteer tourism gap-year industry, argues that the volunteer experience in formative gap-year programmes is designed to aid 'the development of an active, self-regulating, competitive citizen'. Such experiences are increasingly considered necessary starting points for corporate careers because they produce selves with valued competencies (Simpson 2005). The qualities that a good self is meant to embody, such as leadership and teamwork skills (ibid.), are corporatised and marketable versions of the good qualities of any other kind of business or enterprise. The development tourist is produced and subjectified as an 'entrepreneur of himself' (Foucault 2010: 226), through activities in which they become their own capital because they have the abilities and skills that generate income; what Foucault terms 'capital-ability' (2010: 225). The self therefore becomes something that can be invested in and improved economically. The neoliberal economy encourages educational investments as a voluntary increase in human capital (Foucault 2010).

As more educational institutions incorporate volunteering experiences into their curricula, and as programmes are increasingly tied into an educational 'language of "graduation" and "success"' (Simpson 2005: 451), development tourists are given to believe that they are investing in a highly competitive self. Many volunteering companies are collaborating with educational institutions to capitalise on the promise of a competitive self and global development. For example, the Australian volunteering company Antipodeans Abroad has affiliations with many schools and universities, with which it runs regular programmes for students. These programmes are often seen and marketed as a challenging and empowering training exercise for future professionals.

In an article on the gap-year and global citizenship, Lyons *et al.* (2012) argue that the increasing incorporation of gap-year volunteering programmes into higher education pathways is due to the widespread perception that these programmes help form responsible, global citizens. Twenty-one-year-old Teresa completed a development tourism programme in health sciences through a collaboration between the Australian company, Projects Abroad, and her university. Teresa found that she was able to recount her volunteering experiences in resumés as evidence that she is a capable and compassionate citizen of the world. Teresa recalled how

> a couple of weeks ago I had to do a job interview ... and I ended up thinking of things that I could use from Nepal to put in it like, you know, as examples to show my skills ... how I can communicate, and things like that when it's a bit more of a challenge.

Development as ethical transaction

Due to the investment of these tourists in the field, Third World development is increasingly being presented as an enterprise, in which these developing selves 'can offer help, advice and support' (Simpson 2005: 464). For example, employment consultant, Jemma, who did a development tourism placement with the Australian company Antipodeans Abroad in Cambodia, felt that her placement allowed her both to do something different from others and to fully utilise her potential to aid community development in Cambodia. Jemma spoke of the way in which being different is essential to contemporary forms of self-development; as she put it, 'being different actually makes you understand who you really are because you know you stand out from everybody in terms of where you're at and what you're doing'. It was only by volunteering at a school in Cambodia that Jemma felt she could fully develop as a person; as she reflected,

> [A]t home I get frustrated like this is not what I should be doing, not where I'm meant to be, and this is not my full potential as a person. [B]ut when I'm in Cambodia I feel like I'm actually moving towards that full potential.

Volunteers often express dismay or disbelief when development tourism fails to live up to its promise of contributing to a self-as-enterprise, which is sufficiently differentiated from other enterprises as to be marketable. Australian volunteer tourists in the ongoing study answered questions about the growing popularity of volunteer tourism, with a disbelieving: 'Is it popular?', mostly because they used the experience to construct what they believed to be a unique self-as-enterprise. The best volunteering programmes extend the feeling of a unique investment, as Teresa found when she later was able to describe her experience to employers as one in which she had had to 'be independent and stuff and I just explained you know I was around a rural community and in a place where there weren't other tourists'. Demonstrating the wider manifestation of the development tourist's desire for difference, Vodopivec and Jaffe (2011: 117) offer an example of a tourist from their fieldwork in Antigua who was dismayed to find that her attempt to do something good was part of a mass experience: 'I didn't expect so many volunteers. It's unbelievable, it's like a factory.' This dismay is arguably a product of the enterprise society and is derived from the recognition of how common her enterprise-self has been rendered by the mass participation in volunteering and the desire for a more competitive and differentiated investment in herself.

In the enterprise society the goal of development is an increased investment in the self, through the accumulation of human capital rather than the eradication of poverty. The importance of these forms of human capital, such as marketable qualities, skills and experiences, has rendered the self-as-enterprise central to the goals of global development. As Foucault elaborates (2010: 231), in the neoliberal economy innovation and progress comes from the 'income of a certain capital, of human capital, that is to say, of the set of investments that we have made in the level of man himself'. Among the Australian development tourists interviewed, the task of investing in themselves contributed to their privileging of a social enterprise model of development that seeks to expand the ability of all individuals to invest and expand their own human capital. For example, 23-year-old business consultant Jessica, reflecting on the kinds of desirable development that she had learned about while volunteering, stated:

> That is the beauty of the social enterprise, local Africans have the solutions that they need to address the challenge that they are facing it's just a matter of giving them the initial funding in the micro-finance, social enterprise model.

The enterprise rationale behind both the desire among contemporary volunteer tourists for self-improvement and for contributing to poor communities has turned the volunteering field into a formative space for those wishing to shape the future of global development. Eugenia Wickens' study of volunteer tourism in Nepal highlights the contiguity between wanting to be involved in the field of development and the desire for personal development in this field. As various participants in Wickens's study reflected, volunteer tourism in Nepal was motivated by the wish to 'expand themselves personally, while helping others in need' (Wickens 2011: 46). Other participants in the same study were more explicit about their interest in development as a field; as one observed: 'I am interested in development work ... and because it is often said that Nepal is the poorest country in the world, and so my help will hopefully be worthwhile' (ibid.). Wickens's study participants also reflected how this sort of self and societal development fits easily into legitimate, institutional career paths; as one participant described it:

> I had four months free during the summer in between my 2nd and 3rd year at uni....I have always wanted to visit Nepal....I saw an opportunity to go and teach there, but also help in a hostel when the charity was advertised at uni.
>
> (ibid.)

The wide and varied choices offered to development tourists demonstrate a form of self-as-enterprise that is reliant on the maximisation of different interests, motivations and desires. Foucault theorises (2010) that in a neoliberal economy, the multiplication of enterprise forms in society makes individuals' choices between various alternative ends the most significant influence on societal development. In a way that 'absolutises a certain notion of economic interest or choice, [the self-as-enterprise] is never called upon to renounce its interests, indeed they must be pursued and intensified for society to flourish' (McNay 2009: 61). The multiple interests of the development tourists are demonstrated by the sheer variety of programmes offered to them, from short-term to long-term placements, from educational programmes to those which are a minor addition to a traditional holiday; the programmes number in the thousands and continue to expand (Callanan and Thomas 2005; Lyons *et al.* 2009; Benson and Siebert 2009; Ingram 2011).

Development tourism companies are increasingly catering for, and creating a demand for, a highly diversified product, structured by the multiple interests of the enterprise economy. Volunteers can now find a volunteering programme in any glamorous location they could possibly choose. As 22-year-old Katie reflected on her choice of the Maldives as a location for a volunteering project:

> I chose the venue and ... the idea of being able to go to the Maldives, you know, I wouldn't be able to afford to stay on their resort islands or anything like that so the opportunity to go there was a must really.

The programme that Katie chose was in its first year of operation when she decided to take part in it, and is in a glamorous destination that is usually associated with luxury resort tourism. The sheer variety of choice available to development tourists is evident on websites such as that of i-to-i Volunteers International, which specialises in volunteering and English teaching placements across a range of countries; programmes include teaching English to underprivileged children in India, combining a safari with community work in Tanzania, or volunteering for panda conservation in China (i-to-i 2012). On the i-to-i website, prospective volunteers are invited to take a volunteer test to determine the kinds of qualities they possess, or would like to possess, and the experiences for which these qualities are best suited. These tests effectively encourage volunteers to choose a self that reflects one of many diverse development tourism enterprises (i-to-i 2012).

Unsettled accounts

Foucault argued (2010) that mobility was one of the privileged demonstrations of high human capital. Development tourists who, either knowingly or unknowingly, undertake their trips as a way of investing in their self-as-enterprise, accept the inequalities of the neoliberal economy simply by accepting the freedom of mobility within it. Their acceptance is a part of 'a political consensus, inasmuch as they accept this economic game of freedom' (ibid.). Few of the volunteer tourists interviewed expressed a realisation that their mobility itself was a privilege, although all reflected on their relative privileges of affluence and comfort in relation to the communities in which they were volunteering. Often their decision to go to their chosen destinations was a demonstration of their easy mobility and unfettered access to any continent or country in the world. For example, 22-year-old Katie, who volunteered in the Maldives, reflected that while she had wanted to spend the university holidays in Europe, after finding no one to travel with and having heard about the volunteering opportunity, signed up to go 'within twenty-four hours'.

While development tourists may have mobility and be implicated in the wider multiplication of interests in the market, their acceptance of the inequalities of the neoliberal system is contingent on onerous responsibilities. The diversity of the enterprise society reaches a compromise with the widespread inequalities of the neoliberal economy by rendering the self-as-enterprise uniquely responsible for at least a part of the political accountability and governance offered by traditional development institutions. While development has traditionally been the task of states and development institutions, in an enterprise society the efforts of individuals are more differentiated and competitive than the lengthy and heavily regulated programmes of institutional and state actors. This process of what has been called 'responsibilisation', renders global welfare the domain of the market and displaces the responsibility for maintaining it to self-sustaining and autonomous actors rather than governments or institutions (Shamir 2008).

The accounts and reflections of development tourists highlight the substantial problems that arise when the responsibility for global development is placed on individuals. One of the participants in Wickens' study expresses disquiet at the extent of implied responsibility for improving developing communities in the volunteering experience, stating that, 'it has made me more culturally aware and adaptable, possibly more independent as well; it has been enjoyable – though also frustrating as there is so much that needs doing' (Wickens 2011: 48). The frustration that volunteers feel is often compounded by locals expecting young volunteers to have qualifications and experience that they do not possess (Guttentag 2009; Palacios 2010). Often the capabilities gained through development tourism are so vague or ill-defined as to contribute to a feeling of general helplessness. Angela Benson and Nicole Seibert's study (2009: 309) of German volunteer tourists in South Africa offers another example of participants feeling disturbed to 'see that children take glue before they go to school and are on drugs and you can't do anything'. The frustration and helplessness that volunteers feel highlights the ambiguity of their new responsibilities in the field of global development (Simpson 2005). As Vodopivec and Jaffe (2011: 121) state:

Previous generations may have been able to travel unburdened by the pressure to do good. However, many members of the current generation of young people are caught between, on the one hand, this sense of personal responsibility to make a change and, on the other, the structural difficulties of doing so.

Given such a contradictory impasse, development tourists in the enterprise society may exhaust their resources, time and effort in trying to make investments in selves that can barely fulfil the mandate and responsibility given to them.

Conclusion

Despite the difficulties, development tourists are doing their best to live up to the responsibilities imposed on them as enterprise subjects. Cassie and Jessica use the abilities developed through volunteering to give their time to a long list of organisations and services in their local communities. While Cassie hopes to one day work in the field of global development, Jessica volunteers her time with local charities to improve their deployment of the social enterprise model. Jemma plans to extend the fulfilment of the potential and skills that she developed through volunteering to set up a sponsorship programme to support post-school vocational training and higher education studies for young Cambodians. Teresa continues to utilise the skills that she learned on her health science placement at work, and Katie uses the skills from her teaching placement in a new job as a primary school teacher. Both look forward to volunteering again at another beautiful destination when they have developed their skills even further. Most development tourists feel changed for the better through their experiences and usually more so for being able to combine an altruistic contribution to global change with the necessity for self-development at a young age.

Conceptions of what motivates the creation of good selves and a good society within tourism research need to keep moving, at least as fast as the enterprise society transforms these self and global developments into successful enterprises. At the very least, the pathos of development tourists' highly pressured creation of a good self must be appreciated. This appreciation is imperative in order to complicate the understanding of motivations between altruism and egoism in research on ethical tourism. Research into contemporary tourism arguably must take on the responsibility not just of recommending improvements to diversifying tourism enterprises but also that of critiquing the modes of subjectification characteristic of neoliberalism, as accounts of tourism in other disciplinary areas are beginning to do (Vrasti 2013). The latter critique involves the elucidation of possibilities for more equal developments across the world (Mostafenezhad 2012). A detailed deconstruction of what it means to be altruistic or good is increasingly necessary (Godfrey and Wearing 2012) at the intersection of tourism and development studies. Exploring the question of whether the definition of the moral good achieved through development is at all influenced by volunteers' dedicated makings of a self that is good enough to survive in a neoliberal, enterprise society may lead to a deeper understanding of how global (self) development is being sold. To recognise that the global (self) development of the development tourist conforms to neoliberal dictates is not to give up on goals of resisting the neoliberal perpetuation of inequality, goals that volunteers ostensibly share, but rather to demand more reflexive, critical and imaginative visions of social change in tourism.

Activity

Swapping stories

Often the changes wrought by the increasing commodification of self-development in the enterprise society, which are evident in development tourism, can best be seen in our own lives. Before class, divide the students up into groups. Ask everyone to think of a formative or worthwhile overseas trip and to bring in their favourite photo and favourite story from that trip to the next class. In class, ask everyone to share those stories with their group. Then each group can build up a profile of the kind of qualities, personal characteristics or abilities that these travel stories and photos reflect, while considering the following questions:

- How marketable is the kind of person reflected in each group's profile. Do your stories reflect the contemporary enterprise society?
- Have you ever used this or any other experience of travel as a part of a job application or resumé?
- How does the growth of the self-as-enterprise enable us to think differently about the market for ethical products and what is being sold?

Suggested reading

McNay, L. (2009) 'Self as Enterprise: Dilemmas of Control and Resistance in Foucault's The Birth of Biopolitics' in *Theory Culture Society*, 26 (6): 55–77.
Vrasti, W. (2013) *Volunteer Tourism in the Global South: Giving Back in Neoliberal Times*, London: Routledge.
Vodopivec, B. and Rivke, J. (2011) 'Save the World in a Week: Volunteer Tourism, Development and Difference' in *European Journal of Development Research*, 23 (1): 111–28.

References

Barnett, C., Cloke, P., Clarke, N. and Malpass, A. (2005) 'Consuming Ethics: Articulating the Subjects and Spaces of Ethical Consumption' in *Antipode*, 37 (1): 23–45.
Benson, A. and Seibert, N. (2009) 'Volunteer Tourism: Motivations Of German Participants In South Africa' in *Annals of Leisure Research*, 12 (3–4): 295–314.
Boluk, K.A. (2011) 'Fair Trade Tourism South Africa: Consumer Virtue or Moral Selving' in *Journal of Ecotourism*, 10 (3): 235–49.
Brown, S. (2005) 'Travelling with a Purpose: Understanding the Motives and Benefits of Volunteer Vacationers' in *Current Issues in Tourism*, 8 (6): 479–96.

Butcher, J. (2003). *The Moralisation of Tourism: Sun, Sand … and Saving the World?*, New York, London: Routledge.

Butcher, J. and Smith, P. (2010) '"Making a difference": volunteer tourism and development' in *Tourism Recreation Research*, 35 (1): 27–36.

Callanan, M. and Thomas, S. (2005) 'Volunteer Tourism: Deconstructing Volunteer Activities Within A Dynamic Environment' in M. Novelli (ed.), *Niche Tourism: Contemporary Issues And Trends*, New York: Elsevier: 183–200.

Cameron, J. and Haanstra, A. (2008) 'Development Made Sexy: How It Happened And What It Means' in *Third World Quarterly*, 29 (8): 1475–89.

Coghlan, A. and Fennell, B. (2009) 'Myth Or Substance: An Examination Of Altruism As The Basis Of Volunteer Tourism' in *Annals of Leisure Research*, 12 (3–4): 377–402.

Figueroa, R.M. and Waitt, G. (2010) 'The Moral Terrains Of Ecotourism And The Ethics Of Consumption' in T. Lewis and E. Potter (eds), *Ethical Consumption: A Critical Introduction*, London, New York: Routledge: 260–74.

Fornet-Betancourt, R., Becker, H., Gomez-Müller, A. and Gauthier, J.D. (1987) 'The Ethic Of Care For The Self As A Practice Of Freedom: An Interview With Michel Foucault on January 20, 1984' in *Philosophy & Social Criticism*, 12 (2–3): 112–31.

Foucault, M. (2008) *The Birth of Biopolitics: Lectures at the College De France 1978–1979*, translated by G. Burchell, Basingstoke, New York: Palgrave Macmillan.

Foucault, M. (2010) *The Birth of Biopolitics: Lectures at the College De France 1978–1979*, translated by G. Burchell, (2nd edition) Basingstoke, New York: Palgrave Macmillan.

Godfrey, J. and Wearing, S. (2012). 'Can Volunteer Tourism Be More Than Just The Successful Commodification Of Altruism' in *CAUTHE 2012: The New Golden Age of Tourism and Hospitality – Book 2 The Proceedings of the 22nd Annual Conference*, Melbourne, Victoria: La Trobe University Press. Online, available at: www.search.informit.com.au/documentSummary;dn=225187699860357; res=IELBUS (accessed 7 January 2013).

Gray, N.J. and Campbell, L.M. (2007) 'A Decommodified Experience? Exploring Aesthetic, Economic and Ethical Values for Volunteer Ecotourism in Costa Rica' in *Journal of Sustainable Tourism*, 15 (5): 463–82.

Guttentag, D.A. (2009) 'The Possible Negative Impacts of Volunteer Tourism' in *International Journal of Tourism Research*, 11 (6): 537–51.

i-to-i. (2012) i-to-i Volunteering website. Online, available at: www.i-to-i.com/volunteer (accessed 20 January 2012).

Ingram, J. (2011) 'Volunteer Tourism: How Do We Know It Is "Making A Difference"?' in Benson, A.M. (ed.), *Volunteer Tourism: Theoretical Frameworks And Practical Applications*, London, New York: Routledge: 211–22.

Littler, J. (2009) *Radical Consumption: Shopping for Change*, Berkshire, New York: Open University Press.

Lyons, K.D. and Wearing, S. (2008) 'All For A Good Cause? The Blurred Boundaries Of Volunteering And Tourism' in K.D. Lyons and S. Wearing (eds), *Journeys of Discovery in Volunteer Tourism: International Case Study Perspectives*, Cambridge, MA: CAB International: 147–54.

Lyons, K.D, Wearing, S. and Benson, A. (2009) 'Introduction to the special issue on volunteer tourism' in *Annals of Leisure Research*, 12 (3/4): 269–71.

Lyons, K.D., Wearing, S., Hanley, J. and Neil, J. (2011) 'Gap Year Volunteer Tourism: Myths of Global Citizenship?' in *Annals of Tourism Research*, 39 (1): 361–78.

Matthews, A. (2008) 'Negotiated Selves: Exploring the Impact of Local-Global Interactions on Young Volunteer Travellers' in K.D. Lyons and S. Wearing (eds), *Journeys of Discovery in Volunteer Tourism: International Case Study Perspectives*, Cambridge, MA: CAB International: 101–17.

McNay, L. (2009) 'Self as Enterprise: Dilemmas of Control and Resistance in Foucault's The Birth of Biopolitics' in *Theory Culture Society*, 26 (6): 55–77.

Mostafanezhad, M. (2012) 'The Geography of Compassion in Volunteer Tourism' in *Tourism Geographies iFirst*. Online, available at: http://dx.doi.org/10.1080/14616688.2012.675579 (accessed 25 April 2012).

Mustonen, P. (2007) 'Volunteer Tourism – Altruism or Mere Tourism?' in *Anatolia*, 18 (1): 97–115.

Ooi, N. and Laing, J.H. (2010) 'Backpacker Tourism: Sustainable And Purposeful? Investigating The Overlap Between Backpacker Tourism And Volunteer Tourism Motivations' in *Journal of Sustainable Tourism*, 18 (2): 191–206.

Palacios, C.M. (2010) 'Volunteer Tourism, Development And Education In A Postcolonial World: Conceiving Global Connections Beyond Aid' in *Journal of Sustainable Tourism*, 18 (7): 861–78.

Scrase, T.J. (2010) 'Fair Trade In Cyberspace: The Commodification Of Poverty And The Marketing Of Handicrafts On The Internet' in T. Lewis and E. Potter (eds), *Ethical Consumption: A Critical Introduction*, London: Routledge: 54–70.

Shamir, R. (2008) 'The Age Of Responsibilization: On Market-Embedded Morality' in *Economy and Societ*, 37 (1): 1–19.

Simpson, K. (2004) '"Doing Development": The Gap Year, Volunteer Tourists and a Popular Practice of Development' in *Journal of International Development*, 16 (7): 681–92.

Simpson, K. (2005) 'Dropping Out or Signing Up? The Professionalisation of Youth Travel' in *Antipode*, 37 (3): 447–69.

Sin, H.L. (2009) 'Volunteer Tourism – "Involve Me And I Will Learn"', *Annals of Tourism Research* 36 (3):480–501.

Spencer, R. (2010) *Development Tourism: Lessons from Cuba*, Farham, Burlington: Ashgate.

Stoddart, H. and Rogerson, C.M. (2004) 'Volunteer Tourism: The Case Of Habitat For Humanity South Africa' in *Geojournal*, 60 (3): 311–18.

Tomazos, K. and Butler, R. (2012) 'Volunteer Tourists In The Field: A Question Of Balance' in *Tourism Management*, 33 (1): 177–87.

Tomazos, K. and Butler, R. (2010) 'The Volunteer Tourist As "Hero"' in *Current Issues in Tourism*, 13 (4): 363–80.

Varul, M.Z. (2009) 'Ethical Selving In Cultural Contexts: Fairtrade Consumption As An Everyday Ethical Practice In The UK And Germany' in *International Journal of Consumer Studies*, 33 (2): 183–9.

Varul, M.Z. (2008) 'Consuming The Campesino: Fair Trade Marketing Between Recognition And Romantic Commodification' in *Cultural Studies*, 22 (5): 654–79.

Vodopivec, B. and Jaffe, R. (2011) 'Save the World in a Week: Volunteer Tourism, Development and Difference' in *European Journal of Development Research*, 23 (1): 111 28.

Wickens, E. (2011) 'Journeys Of The Self: Volunteer Tourists In Nepal' in A.M. Benson (ed.), *Volunteer Tourism: Theoretical Frameworks and Practical Applications*, London: Routledge: 42–52.

9 Exploring the ethical discourses presented by volunteer tourists

Karla Boluk and Vania Ranjbar

Introduction

Travellers often demonstrate a variety of motivations that propel them to make their travel-related decisions. Taking time out of one's everyday life, relaxing, seeking sun, pursuing adventure, experiencing new environments, education through interacting with different cultures, learning or practicing new languages, eating new foods, smelling new smells, taking in different scenery, and visiting internationally renowned places may comprise just a few of the motivations drawing people to travel. The media can reinforce such wishes. Given the array of possible motivations highlighted, it can be said that some travellers may be attracted by several motivations at once. Alternatively, one or more motivations may take priority for part of one's holiday, which may be replaced at a later time with other motivations. For example, an individual travelling to China may initially be motivated to take up the Chinese language and engage in a process of cultural immersion, while in another part of their holiday they may be motivated to indulge in Chinese cuisine. Clearly, travellers' interests and motivations can be diverse, and may change throughout the course of their holiday.

Using specific discourses before, during and upon returning from one's holiday may further incentivise one's specific travel-related decisions. Such discourses may add stimulus to specific ethical travel choices. Thus, travellers enacting their motivations, exploiting specific discourses, and being *seen* in a particular light, may factor into the decision to participate in particular activities for a part of, or all of their holiday. Therefore, conspicuously demonstrating their *intent* to their significant reference groups may incentivise choosing particular ethical holidays and/or lend the opportunity to describe their participation in a specific and opportunistic way. For instance, a traveller concerned with investigating how they could augment their emissions, may be interested in not only paying attention to environmental impacts but also appearing well educated and informed. Consequently, the discourse used to highlight one's participation may be synonymous with environmental awareness, global citizenship, being part of the collective, bigger picture, and/or demonstrating leadership capabilities. Similarly, those keen to travel for the purpose of volunteering may choose for myriad reasons to prioritise certain discourses in an effort to situate themselves as well intentioned, altruistic and self-sacrificing. Such discourses may overpower other more hedonic motivations such as experiencing a new culture, seeing a new part of the world, practicing one's language skills, and building one's CV. As such, one is able to demonstrate one's good self via their tourism decisions and they may choose to prioritise the former over the latter.

The aim of this chapter is to explore how various discourses are used and how volunteer travellers perceive their volunteer work at Cotlands, a non-profit organisation in South Africa. The paper presents data from nine in-depth interviews that were carried out with volunteers between 2010 and 2011, and utilises discourse analysis to examine the various meanings of participants' experiences. Research in the area of volunteer tourism prioritises the various motivations that encourage participants to choose this type of holiday. However, there is currently a dearth of research exploring the various interpretations and meanings of such motivations.

Volunteer tourism

Volunteer tourism refers to those who 'volunteer in an organised way to undertake holidays that might involve aiding or alleviating the material poverty of some groups in society, the restoration of certain environments or research into aspects of society or environment' (Wearing 2001: 1). Interestingly, Wearing's oft-cited definition seemingly neglects the question of 'what is in it?' for the traveller. In Singh and Singh's (2004: 184) work they argue that volunteer tourism is a 'more conscientized practice of righteous tourism – one that comes closest to utopia', and in its best form upholds 'the highest ideals intrinsically interwoven in the tourism phenomenon'. Similar to Wearing's definition, Singh and Singh centralise the importance of the 'altruistic' and 'giving' aspects of travellers. More specifically, volunteer travellers will make a tangible contribution to the communities they visit. McGehee and Santos (2005: 774–5) regard volunteer tourism as an 'opportunity to provide ways to create and establish relationships' beyond transitory encounters, in addition to 'enhancing the notion of "global citizenship"'. Again, McGehee and Santos's definition recognises an outward focus. Perhaps more realistically, Callanan and Thomas (2004: 183) highlight that volunteer tourism has a clear and twofold focus that considers altruism and 'self-developmental experiences that participants can gain during their time working on such projects'. This paper argues that volunteer tourism, for some travellers, may significantly encompass a third of attraction; specifically, that volunteer travellers find appeal in the notion of constructing their self-image, in light of their self-sacrifice, to their significant reference groups. The question framing our study is this: do volunteer tourists use their holidays to facilitate their self-construct to specific reference groups?

Despite the notion of volunteer tourism as virtuous, some researchers (e.g. Bennett 2008; Sin 2009) have begun to question whether it is effectively equivalent to other benevolent forms of tourism (e.g. 'just tourism', Pro Poor tourism, Fair Trade tourism) that contribute value to the communities they originally sought to assist. Sin's (2009: 497) study found that common motivations among the 11 volunteer tourists he interviewed were 'to travel rather than to contribute or volunteer', clearly challenging the definitions put forth by Wearing (2001) and Singh and Singh (2004), who primarily focus on the benign aspects of volunteer tourism and the altruistic characteristics of the volunteer tourists themselves. In conflict with the definitions mentioned above, some researchers argue that volunteer tourism is a form of post-colonialism. For example, Dhruvarajan (2000) purports that volunteer tourism surfaces as an additional exploitative form of tourism occurring in the Majority World. Although volunteer tourists may, in some circumstances, refrain from making the same sort of environmental and economic impacts caused by participation in 'mass' tourism, a social impact may be unavoidable.

Research in the area of volunteer tourism is currently in its infancy. While contemporary scholarship primarily focuses on the motivations of volunteer travellers, there is little research to date that explores the notion of constructing one's self-image and/or ethical self within the context of volunteering abroad, all of which which add another dimension to the motivations propelling individuals to engage in certain activities while on holiday. Volunteering freely in one's own community is perceived as favourable by many societies; however, the notion of volunteering while on holiday may encompass different meanings because in most cases people pay to volunteer. Paying to volunteer may give travellers a platform from which to distance and distinguish themselves from their peers, something that may be done unconsciously or intentionally.

Volunteer tourism as an opportunity to develop oneself

In addition to the benefits created for communities, volunteer tourism experiences can also provide an opportunity to develop one's sense of self. Thus, the contributions to a particular society can be secondary to personal development and fulfilling objectives relating to the self (Sin 2009). Such an inward focus pushes aside the altruistic perceptions often affiliated with volunteer tourists, as depicted in the academic definitions previously cited. Volunteer tourists are then differentiated from their mass tourist counterparts as they participate in an 'alternative' form of tourism.

The central discourse of volunteer tourism is that it is an admirable act involving a level of commitment. This is explained by Mustonen (2005: 171), when he says that 'only those tourists, who are seriously motivated by other than purely touristic factors, travel to these destinations', the 'these' referring to the fact that most volunteer destinations are located in the Majority World (see Higgins-Desbiolles 2006: 1202; Shaw and Newholm 2002: 174–9), where resources are strained and not necessarily supporting of luxury and carefree holidays. Such destinations may be remote and difficult to access, which contributes additional barriers. Accordingly, there is a discussion of the 'liminoid', positioning volunteering as a rite of passage (Mustonen 2005) which may be the case for young people seeking opportunities during their gap year as a way to be productive and enhance their competitive advantage (often displayed on their CV) while simultaneously taking a break from study or work commitments. In the context of the liminoid discussion volunteer tourism may provide an opportunity for some individuals to engage in a transformation from everyday life as a consequence of physical movement, as pointed out by Mustonen (2005). This separation from one's social groups and situations may allow an individual both time and space to reinvent themselves before they are reintegrated into their home life.

Research suggests the benefits created by volunteer tourism for the individual volunteer include self-gratification (Guttentag 2009), mental and physical healing from corporate burnout (Brown 2005), an enhanced sense of self, and more specifically a sense of physical, emotional and spiritual fulfilment from participation (Stoddart and Rogerson 2004). Furthermore, the significance of informal qualifications has been noted (Sin 2009), as well as opportunities for CV development (Simpson 2004), a factor that has been particularly attractive in propelling gap year volunteers. As such, a number of desired outcomes are apparent in volunteer tourism such as altruism, personal development and finding one's way (Sin 2009). Ultimately, volunteer tourism is presented as an opportunity for those at a critical time of change in their lives, more specifically those who are in a state of being 'betwixt and between'. Accordingly, for some, volunteering may represent the next logical step (McGehee and Santos 2005: 774).

Maximising the skills of volunteers with the needs in communities has warranted some attention (see Barbieri *et al.* 2012). Failing to recognise the skills of volunteers in the assignment of specific tasks can lead to feelings of disappointment and/or a sense of obligation on behalf of the volunteer. Thus, communication between the organisation and its volunteers is key if the skills and interests of volunteers are to be effectively matched (Guttentag 2009) in order to achieve desirable outcomes.

Discussions regarding the development of self subsequent to volunteer experiences have gained some interest in the academic discourse. A stronger sense of self has been established as a consequence of volunteer work; notably personal and interpersonal awareness, self-contentment (Wearing 2002), 'self-actualization, self-enrichment, self-expression, recreation or renewal of self, feelings of accomplishment, enhancement of self-image, social interaction and belonging' (Stebbins 1992: 6–7), increased cultural understanding, social awareness, sense of global responsibility (Morgan 2009), self-reflection, value or belief change following their experiences (Singh and Singh 2004; Stoddart and Rogerson 2004; McGehee and Santos 2005) and life changing or cathartic experiences (Zahra and McIntosh 2007). Such effects have a positive impact on host organisations and projects, as they often feel as though they are given a voice (Morgan 2009). As such, volunteer experiences have the potential to derive meaning, and may foster a process of co-transformation through social exchange, preparing volunteers to navigate a new social world as a result of their experiences. What the literature on volunteer tourism has not investigated to date is specifically how travellers perceive their volunteer work, and the various discourses that they use to potentially demonstrate their ethical selves.

Volunteer tourism as an opportunity for ethical selving

Volunteer tourism has been presented as an opportunity to participate and expand activist identities, thereby enacting one's personal beliefs through the medium of political activism (McGehee and Santos 2005). This is similar to the decision some consumers make to 'buycott' Fair Trade-certified products as a result of their supplier or producer's environmental and social justice commitments. Such political activism during one's holiday may present an opportunity to consolidate identities (Sin 2009: 487), 'to choose, to be consistent' between one's home life and travelling. An inherent dichotomy exists for some traveller volunteers between the tourism element of their holiday, their altruism, and a search for fun (ibid.). These contrasting elements demonstrate a dual focus on one's self and others (Sin 2009). As such, a combination of hedonism via tourism appears as an important component of the traveller's choice to volunteer and their altruistic consideration of others. But what happens when hedonism takes place within the same environment as volunteering, spilling over into the volunteer tourism arena?

Social norms and expectations play a significant role in what and how individuals choose to consume (Andorfer and Liebe 2013). As such, consumers can gain more than utility from the products that they purchase. Identity dimensions in the process of ethical consumption, such as purchasing Fair Trade coffee, choosing a Fair Trade holiday, or engaging in volunteer work, are associated with an ideal social image (Andorfer and Liebe 2013). Therefore, such consumption not only supports small-scale producers (and/or assists in the area of development projects) but also expresses one's consumer identity. Consumers can construct, categorise, reinforce (Varul 2009) and express the identity they aspire to as a consequence of their purchasing behaviour (Barnett *et al.* 2005).

Ethical consumers, by enacting political and moral concerns via their consumer choice, are simultaneously constructing themselves as ethical (Varul 2009). Consumers illustrate their agency and accountability in the function of their social interaction (Foucault 1987). Ethical consumption and the consumption of Fair Trade Tourism and volunteer tourism, for example, are embedded into a cultural context of global consumer capitalism (Billig 1999). This macro discourse informs the way people think about the extent of their responsibility, what constitutes a fair exchange, and how they construct themselves as ethical consumers. Ethical consumption can be described as a form of conspicuous consumption because consumption, in itself, is a hedonistic act (Holbrook and Hirschman 1982; Schaefer and Crane 2005), especially when used as a mechanism that lends visual expression of an 'ethical self' to the world. The consumption of volunteer tourism provides similar opportunities for such a visual demonstration, and may allow individuals to appear as though they are even more consistent because they are demonstrating that their ethical concerns also influence their travel behaviour. This may be increasingly significant for individuals, and their reference groups, given that consumption in the context of tourism has been recognised as hedonistic. It follows that travellers who seek out ethical holidays may appear to be exceptional.

Methodology

Data collection

In 2010 and 2011 ten semi-structured interviews were carried out with volunteers during their time at Cotlands in South Africa. Semi-structured interviews allow informants flexibility to influence the course of the conversation and uncover the significance and meaning of particular topics. This data collection method was chosen in an effort to gain rich insights from volunteers regarding their perceptions and the personal meanings they had taken from their volunteer work.

Cotlands is a non-profit organisation in South Africa caring for vulnerable children up to 12 years old (and their families), in residential and community care (outreach) programmes focusing on three main areas: health, education and psycho-social. Services are provided nationally in six regions, with the headquarters based in Johannesburg, Gauteng. At the time, Cotlands had approximately 200 employees nationally. Initially, only Gauteng was able to accommodate international volunteers; however, in 2011, Cotlands' Western Cape project was able to accomodate a small number of international volunteers. The interviews for this study were carried out with volunteers in Gauteng. Volunteers assist staff in the daily operation of various programmes, as well as carrying out administrative chores.

Participants

Nine interviews were carried out with seven females and two males who were volunteering at Cotlands for periods of between one month and one year. Specifically, one informant volunteered for a month, six volunteered for three to six months, and two of the nine volunteered for a year. Those taking part ranged in age from 19 to 42. Informants were from Ireland (one), Germany (two), UK (two), the Netherlands (one), Switzerland (one) and Australia (two). A majority (five) of the particpants were in a gap year, while the remaining four were taking time off from their careers to volunteer. Pseudonyms were used to protect the identity of the respondents.

Analytic technique

Discourse analysis was chosen for this study. Discourse analysis deals with socially constructed frameworks of meaning, which contain information regarding what an individual deems to be or not be acceptable (Foucault 1972). Discourse, then, is influenced by, and located within, wider social structures and influences how individuals construct their language and ultimately present themselves. Discourse analysis 'focuses on quintessentially psychological activities, activities of justification, rationalization, categorization, attribution, making sense, naming, blaming and identifying' (Wetherell and Potter 1992: 2). Thus, for discourse analysts, language is not a means for merely describing a phenomenon; language is a means for doing something – language *is* action (McKinlay and McVittie 2008). Discourse analysts systematically 'decode' (Taylor 2001) discourse using social and cultural knowledge obtained via membership of social and cultural groups. The task at hand, then, for the analyst, is to understand how versions of reality are constructed to perform action; that is, to understand the process underlying, for example, the production of certain identities during discourse, as well as their practical consequences (McKinlay and McVittie 2008). For discourse analysts, identities can be seen in terms of different 'positions' that speakers take or avoid taking during the course of a conversation (ibid.). The aim of employing discourse analysis in this chapter was to explore the meaning informants associated with their time volunteering at Cotlands and if/how their experiences played a role in their self-construct. In our analysis we paid attention to how language was used, its variability of use and meanings derived, and outcomes of a particular discourse as per guidelines provided by well-established discursive researchers (e.g. Antaki *et al.* 2003; Wiggins and Potter 2008).

Results

Four themes emerged from the analysis: (1) personal fulfilment, (2) detachment and reflections of returning home, (3) distancing discourse, and (4) preferred ways to contribute. Each theme will be discussed below.

Theme 1: personal fulfilment

Engaging in volunteer work and participating in a volunteering community was a significant aspect of the volunteer experience. Volunteer engagement and participation was constructed as not only fulfilling, but also personally rewarding for those involved. The rewards and fulfilment derived from volunteering at Cotlands was described in three ways including: directly working with children and acquiring new capabilities (and skills); witnessing and engaging with hardship; and escaping day-to-day stressors.

For many informants interacting with the children led to fulfilling experiences. Specifically, four informants mentioned this. For example, Maeve (Ireland, aged 42) described a sense of fulfilment when she said this:

> You walk into a classroom and they [the children] start shouting your name....I was holding a baby there the other day and the baby fell asleep....If someone had offered me money at that stage for the baby not to be on my knee I know what I would've taken and it wouldn't have been the money.

This quotation exemplifies Maeve deriving fulfilment from her experience volunteering and her time spent with the babies. Another informant similarly reflected on the personal benefits derived from his experiences of volunteering. Jake (German, aged 19) said 'when a child is crying and you just hold him [and/

or] feed him ... or they fall asleep in your arms ... its relaxing....I think that is nice....I can take some good experiences home'. Ultimately, the work described made the informants feel good. Both examples highlighted that they felt useful and needed, which are often incentives for volunteers (Bussell and Forbes 2002). Such moments encouraged both informants mentioned above to reflect on what is important in life. Significantly, in Maeve's example, money was placed in contrast to working with children with HIV/AIDS. This may have been a way to demonstrate her ethical self by demonstrating that money and possessions were losing significance; the giving of her time was prioritised as a more meaningful transaction.

Such discourses could suggest that 'volunteering' may benefit the volunteers more than the receiving community or the host organisation, supporting the desire for both informal (Sin 2009) and formal qualifications (Simpson 2004) sought by volunteers. The sense of fulfilment demonstrated by both informants reveals the beneficial and meaningful opportunities that can be created by volunteering on one's holiday. The importance of self-reflection was commonly shared in both discourses, as well as the significance of volunteering for one's self-awareness and thus self-construct.

Acquiring new capabilities through working with the children also led to personal fulfilment. A realisation of the significance of the 'moment' created by volunteering in Cotlands was recognised by two informants as they acquired new skills. One, James (from Switzerland, aged 19), recognised that volunteering in Cotlands 'is a very important moment...This experience is for life....I've never worked with children before and learnt how to manage them, what they need or what they think'. Indicative of James' discourse is that his experience is personally relevant and clearly fulfilling. However, his experience is not one of his volunteering being beneficial to the charity, rather it is framed as beneficial to himself as a result of the experiences he has gained and the things he has learned. Similarly, the experience working with children with disabilities and behavioural challenges was something that Denise (Australia, 29) felt she got out of the experience volunteering. Also important to Denise was that her volunteer work allowed her to effectively act as a 'responsible citizen'.

Further to the personal rewards derived from working with children and acquiring new capabilities, a sense of curiosity emerged as important to some informants. That is, witnessing and engaging with hardship that one had previously been curious about informed respondents' decisions to volunteer and led to a sense of personal satisfaction and fulfilment. For example, one informant reported: 'It's only when you do something like this and you see sick children that it becomes real and I'll go back to Ireland a very fulfilled person' (Maeve). Encountering hardship was also significant for Jake:

> I want to become a doctor so I am very interested in diseases and stuff....I thought it would be great to come to a place like this to come in contact with [the disease] HIV positive. I had never imagined that I could come so close to children [with HIV].

A sense of personal interest is at play in this example; however, an alternative interpretation is that the informant derived benefit from experiencing disease first-hand, thereby confirming his interest in becoming a doctor.

A third example was found in Kathryn's (from Australia, aged 28) discourse: 'I was really drawn to come here and see on a bigger scale ... [what] these kids are dealing with; all of these problems brought on by AIDS which we don't have in Australia so I really like that.' Although this suggests that Kathryn was observing something far removed from her realm of experience in Australia, there is a connotation of virtue as well as curiosity about a place where children are overcome by adversity. Interestingly, what is absent in the discourse is any discussion of addressing needs in an Australian context, which might have suggested a way that Kathryn could contribute to Australian society upon her return.

Last of all, the feeling of fulfilment through volunteering was described, perhaps unconsciously, by one informant. Kathryn put forth that volunteering at Cotlands provided an escape from her day-to-day stressors and bereavements of the previous year. Accordingly, volunteering was a way for her to remove the focus from herself. Specifically, Kathryn explained that

> it's easy to get bogged down in your everyday life....I experienced the death of my father, grandfather, aunt and great aunt in the last twelve months and so we really needed to get out and get some perspective.

Clearly, to Kathryn, volunteering in South Africa was perceived as a way to regain control and/or some perspective after enduring a string of bereavements. Furthermore, volunteering would provide an opportunity to focus outwardly, rather than engage in self-pity.

The complexity of the discourse also suggests that going to volunteer was a chance to figure things out, catch one's breath after an intense year, and regroup. As such, volunteering is used for mental and physical healing (Brown 2005), as a means of creating social interaction and a sense of belonging (Stebbins 1992), and perhaps as a method for emotional and spiritual healing (Stoddart and Rogerson 2004). One might question exactly how individual fulfilment was being achieved by Kathryn. We argue that going abroad removed her from a situation where she was reminded of her losses, and provided a chance to spend time with individuals whose circumstances (poverty, terminal illness) might have been perceived as more dire than her own. Moreover, spending time volunteering might have been perceived as enhancing one's ethical self, appearing to be invincible and able to rise above personal crises while simultaneously prioritising the needs of others.

Interestingly, few informants reflected on or addressed the notion that their time spent at Cotlands might be beneficial to the host organisation or its beneficiaries. This challenges the definitions put forth by Wearing (2001) and Singh and Singh (2004) who centralise the notion of making a contribution as the dominant goal of volunteer tourism. The examples demonstrate the complicated agendas of volunteers and the favoured reasons for volunteering in today's society – assisting in the development of one's self-construct and promoting an ideal self. A similar lack of reflection on the benefits to Cotlands was also noted when informants began to discuss their feelings about leaving.

Theme 2: detachment and reflections on returning home

The second theme emerged from informants' discussions around what it might mean to leave Cotlands after their time volunteering. The notion of detachment was referred to in two ways; initially detachment from the children raised concern, followed by a reflection on the consequences of readjusting to their Western lifestyles.

Spending time with the children and babies at Cotlands was a favoured pastime expressed by many of the volunteers. The difficulty associated with leaving them behind was expressed by some. Sarah (from the UK, aged 19) said: 'Telling the kids this week that we're going, they're like "where are you going?" and it's so touching to see ... you have made an impact on them ... 'cause you are making a difference.' The volunteers are convinced that the Cotlands children do not want to see them leave, but the reality is that they are constantly saying goodbye to volunteers. Thus, the disappointment among the children is a potential consequence of attachment to the transient volunteers. Similarly, Kathryn said:

> I know that upon leaving ... there is actually a significant mourning period [regarding] your contribution here and the very real life and death matters that you deal with....When you go home it's very hard to balance them out [and] go back to normal [...] and be concerned with having to buy a toothbrush when you've dealt with people who don't have anything [...] and mourning the loss of the children that you've grown to love.

In both the examples above the volunteers prioritise the difficulty felt by themselves over the difficulty that will be felt by the children. Prioritising oneself over the children could have been purposeful if the informants recognised that they were developing bonds with some of the children and that they were distressed at the prospect of leaving behind the connections they had made. Alternatively, minimising the needs of the children might have been unintentional given the significance of the experience perceived to have been gained and the magnitude of subsequent changes anticipated as a consequence of volunteering (e.g. McGehee and Santos 2005).

Readjustment to a Western lifestyle was described as difficult by many informants following their volunteer work. However, the focus is significantly placed on oneself enduring a sense of 'loss' rather than consideration for the consequences for the community and/or how one might perceive and contribute to local challenges upon returning home. Barriers connected with the disparity in wealth between the conditions in Cotlands and a Western lifestyle may feed into the development of one's self-construct, and, invariably, one's ethical self

(Varul 2009). Sentiments of 'living rough' for a while and/or being distressed by trying to resolve personal tensions concerning the imbalance of power and resources between Africa and their home countries may position the informants as overburdened agents trying to make positive change.

For some informants it appeared as though a personal transformation took place as a consequence of their time spent volunteering. The realisation of such a transformation caused concerns for volunteers regarding what they might face upon their return home. Maeve said:

> I think it's gonna be very hard to adjust to life again....I think you find almost like a sense of peace when you're over in with the children you know it's everything you do is a bit more worthwhile to what you do at home.

She went on to say that

> everything seems a bit more valued here than it does back where I'm from. So to go back to people just talking about what car they've bought and how much money they're spending ... that's gonna be very hard when I've seen what I've seen here.

A sense of belonging, common in volunteer tourism research (e.g. Bussell and Forbes 2002), was inherent in Maeve's discourse. It is apparent that she believes she is making a worthwhile contribution which may be a consequence of the connection she is making with the community and/or because society has labelled volunteer work as 'good'. The 'sense of peace' that Maeve encountered contrasts with the Western lifestyle that she left behind. Thus, her discussion focused on how it may be difficult to return to her same way of life. Specifically, money and her privileged way of life back home were contrasted with the way of life in South Africa. It seems apparent in Maeve's discourse that she has constructed her time and her work in Cotlands as worthwhile, and perhaps noble. Inherently, she may be constructing herself as ethical because she has given of her time and her energy through her volunteer work, none of which has been easy. This may then distinguish herself from her friends and family back home.

Theme 3: distancing discourse

Some informants' discourses demonstrated a sense of distancing between themselves and those family members, friends and acquaintances that they left back home. Maeve expressed feelings of being 'judged', particularly with regard to a benefit night that she had organised to raise awareness and funding for Cotlands before her visit.

> When I was doing the speech thing someone turned around and said to me why are you doing this? And I said because everyone talks about it and nobody does it.... I'd like to do more.... I think it would do everybody good.

Maeve's discourse separates her own accountability and future actions from the inaction of her peers. Throughout her interview she also seemed to challenge her own lifestyle as well. As such, engaging in charity work was contrasted to her own generally 'glammed up' lifestyle back home (as a sales assistant), as well as those in her community who represent the 'haves' in society. Ultimately, upon reflection, Maeve tried to develop an idea regarding how she could be a better person and a better contributor to society. Perhaps her actions were a way of distancing herself from all those who 'say' but never 'do'.

Similarly, another informant referred to what many of her peers were spending their gap years doing, and their choices were juxtaposed with her decision to volunteer at Cotlands. Sarah said:

> I know a lot of people who are on gap years and it makes me so angry when I hear them saying 'I'm going to South America, I'm gonna get drunk for like six months and then [go] ... to Thailand to do much of the same.'

Although Sarah recognises that it is completely justifiable to have fun during a gap year, she argues that 'you have to give something back otherwise it's completely selfish'.

Highlighting the travel choices of her peers and their alcohol-consuming behaviour was contrasted with telling 'everyone' upon returning home about 'all the stuff that you've been doing' and being rewarded by watching 'their faces light up' as they recite 'wow [it] really does sound amazing I really want to do something like that ... and they are inspired by what you've done'. Sarah's choice to volunteer during her gap year may therefore have been motivated by the responses she received from her peers in the past following her previous volunteer tourism experiences. In addition, volunteering appears superior when contrasted to some of the more hedonistic pursuits sought by some of her peers. This may affect Sarah's self-construct, the way she perceives volunteering, and her perception of her ethical self.

At a personal level, volunteering at Cotlands appeared to be a way of counterbalancing a previous trip Sarah took to South America for leisure purposes; however, Sarah described how she felt about her volunteer work and how her peers perceived her as a consequence of this. Her account suggests that a conscious choice to selflessly spend part of her gap year volunteering gives a descriptive value to her 'good deeds' and perhaps explains her motivation in terms of the satisfaction gained from how she anticipated her peers would react when she returned home and summarised her contribution. Such an interpretation could depict her volunteer work as rather hedonistic, selfish and/or self-fulfilling as she consciously refers to volunteering as exceptional and altruistic which reflects the socially desirable construction in society (Andorfer and Liebe 2013). Such reported findings in volunteer tourism studies are in a minority (Guttentag 2009). In the case of Sarah's interview it seems as though some of the discourses she projected reflect self-worth via peer acceptance, something which is not currently reflected in the literature on volunteer tourism and may be another motivating factor behind volunteer tourism.

For Natasha (from the UK, aged 19), when asked why she decided to volunteer during her gap year she said: '[It] was very important to me.... I didn't want [to be] completely self-indulgent in my holidays.' Natasha went on to identify how her choices appeared exceptional when compared to other options that were available to her. For instance:

> I was very very keen to do a ski season but I think there's a very big danger having an incredibly self-indulgent year out when you can just travel the world [and] do what you want, which for some is fine but ... [not] for me and my parents [who have been] heavily involved with charity work.

This discourse demonstrates Natasha's previous interest in volunteer work, which suggests that fashion and impulse are not her driving force. As a further way of validating her ethically sound behaviour Natasha counterbalances what she could have chosen to do for her entire gap year – that is, spend a season skiing. While Natasha denotes skiing as 'self-indulgent' and hedonistic she actually spent a part of her gap year as a ski instructor in France and one month volunteering at Cotlands. But the dominant discourse has ignored this. Volunteering in Africa is positioned as socially responsible based on her privileged upbringing and global awareness, and hence her actions are visibly making her feel good about herself. This might be interpreted as not being completely altruistic, and perhaps even elitist. Ultimately, volunteering appears to be regarded as an obligation, as evidenced by the suggestion that travelling and 'do[ing] what you want' is 'fine' for some.

Natasha adds during her interview that there may be a 'selfish inclination as to why people volunteer and it may be to make themselves feel better'; however, she quickly distances herself from those individuals, saying: 'I don't think I fall into that category ... because I've done it for so long and it's a routine.' The use of the word 'routine' implies obligation, once again. Thus, Natasha appears sensitive to the notion that volunteering is not always perceived by others as completely altruistic but that it can in fact be seen as selfish; one cannot escape the 'feel good' factor that comes along with doing something good for others. The benefits of volunteer tourism, as derived from volunteers, have taken the focus of the literature (see Stoddart and Rogerson 2004; Brown 2005; McGhee and Santos 2005; Guttentag 2009; Sin 2009); the sense of obligation in the context of volunteer tourism has not been established in the literature to date. But Natasha constructs the work she and her parents do as mundane when she describes volunteering as a 'routine'. This notion, from Natasha's perspective, could position her and her family as thoughtful and good people.

The final theme explores the preferred mechanisms through which some volunteers wanted to contribute.

Theme 4: preferred ways to contribute

A preference about the ways that informants wished to contribute to Cotlands was recognised in some of their discourses. Specifically, one of them demonstrated an intent to acquire specific experiences as a way of supporting his career goals; other informants had a preference for interacting with the children, while one informant described that her commitment to volunteering was limited to her time abroad at Cotlands, due to time constraints back home.

Giving in a certain way was highlighted by one informant (Jake) whose discourse demonstrated an innate concern with 'helping' people in the 'volunteering' environment as a consequence of his interest in becoming a doctor. However, he also displayed behaviour that might be perceived as counterproductive to this aim; on one occasion he purchased a Bafana Bafana (the South African national football team) t-shirt from the 'guy on the corner' for R100 instead of paying R300 in the sports store at the shopping mall. The impression given in his subsequent telling of the story was that the 'guy on the corner' obtained his goods through questionable and perhaps illegal means. Jake would also partake in the type of food-wasting behaviour often displayed by volunteers who would often take much than needed from the food donated to them by a local grocery store, and consequently have to discard the extra. Such inconsistencies between discourse and behaviour may suggest that volunteers valued or prioritised certain aspects of 'helping' more than others; for example, medical help was valued or prioritised while the whole social ethos that one might perceive to be an important aspect of volunteering was, at times, absent.

For many informants spending time with the children was the highlight of their time at Cotlands; accordingly, it was the preferred job for most. Throughout Melissa's (from the Netherlands, aged 23) interview it appeared as though she was disappointed with the tasks assigned to her at Cotlands; it appeared that she thought she should have more input into how her time should be utilised while volunteering. Melissa's discourse prioritised an interest in spending time with the children and was critical of having to spend time carrying out tasks of an administrative nature. As Melissa put it: 'I didn't really come to South Africa to copy for hours.' Later on in the interview she said: 'I like doing admin stuff once in a while but hours of copying; it's just not really what I am here for. So I don't mind doing it but it would be better if they [Cotlands] would change that.' Thus it appears that some volunteers had quite specific ideas about how they might contribute/volunteer, ideas which at times may have been incongruous with the realistic needs of the volunteer organisation. This discourse could be interpreted as meaning that the volunteer concerned was not motivated to volunteer where help was actually needed but rather in specific ways that were self-fulfilling. Furthermore, this example may demonstrate that some volunteers are not interested in being told what to do, rather they are more keen to pick and choose what interests them specifically, and what jobs might match their skills and interests (Guttentag 2009). As such, volunteers can still be ethical, and indeed helpful, but in specified ways.

The significance of a volunteer's contribution is realised within this discourse; specific 'contributions' may reflect different meanings to different volunteers, and certain jobs may undermine an individual's education, experience and capabilities. This may then have an impact on the stories that they are subsequently able to reproduce to their significant reference groups. If the tasks that they are given are boring, this experience might not be something they can share in a way that will win respect from their peers and lend support to the notion of ethical self. Also, because volunteer tourists pay for their experience, informants may be concerned about getting a meaningful return on their investment.

For one informant volunteering offered an opportunity to contribute to her personal social capital during her gap year. Henrike (from Germany, aged 20) said: 'I grew up under really wealthy conditions and I wanted to get to know something different.... I wanted to do something social in my life for a certain time.' Henrike's comments reveal that she wanted to fulfil and perform her share of social responsibility, but perhaps only for a 'certain time' rather than leading anything near a full-time lifestyle that continually engages with social responsibility. Thus, there was a clear end date to her social contribution. Contributing for a limited period of time challenges McGehee and Santos's (2005) assumption that volunteer tourism creates and establishes relationships for the long term. One may interpret curiosity in the discourse; Henrike was keen to experience and try out something different from her more familiar wealthy lifestyle. There was also a sentiment in the discourse that *ticking the social justice box* was important, that a short stint of volunteering might be just enough to position herself as good and ethical to her peers. In this circumstance this informant does little to enhance the notion of global citizenship that McGehee and Santos (2005) argue is the driving force behind volunteering.

Discussion

This paper advances the scarce research on volunteer tourism, and specifically on the notion of constructing one's self-construct and/or ethical self. We argued that volunteer tourism, for some travellers, may significantly encompass this third element, an element which may act as a motivation for travellers in addition to their desire to make a contribution to an organisation or community, or fulfil an aspect of personal development. The paper explored whether volunteer travellers may be attracted to the notion of constructing a self-image in light of the self-sacrificing nature of a volunteering holiday that might be perceived by a their significant reference groups. To this effect, we specifically explored how one's construction of self may bring meaning within the context of volunteering at Cotlands. The question framing this study was 'do volunteer tourists use their holidays to facilitate their self-construct to specific reference groups'?

Four overarching themes emerged from the discourse analysis: personal fulfilment, detachment and reflections on returning home, distancing discourse, and preferred ways to contribute. Each of the themes similarly demonstrated an inward focus by the volunteers, with contributions to Cotlands demonstrated as secondary. This finding is similar to Sin's (2009) study that found personal development and fulfilling objectives relating to the self to be important in volunteer tourism; a primary focus, in fact. This study found that personal development was particularly important as a way of constructing one's 'good' self and this appeared to be a priority. Such inward focusing pushes aside the altruistic perceptions often affiliated with volunteer tourists, as often depicted in academic definitions. This study clearly illustrates that the experiences and benefits volunteers were personally gaining as a consequence of their time away from home, and the community's position (impoverished and in need of help) in allowing them to gain perspective, dominated their discourses.

More importantly, this study found that certain priorities were established in volunteers' discourses in an effort to situate themselves as well-intentioned, altruistic, self-sacrificing, or similar. Furthermore, such discourses seemed to overpower other more hedonic motivations such as experiencing a new culture, seeing a new part of the world, taking time out from everyday stresses, and building a better CV. Informants put forth a number of discourses projecting their ethical selves. For some, their altruistic actions placed them on a pedestal. Another discourse, whether constructed consciously or not, placed volunteering as a priority during her time of personal bereavement, which may have led to notions of invincibility and placing others' needs before her own.

Resolving tensions between the balance of resources in Africa and the West dominated the discourses of some informants, which may have reflected their commitment to trying to reconcile the very serious issues that they witnessed. Informants' discourses also compared and distanced their own actions from those of their peers, which in turn helped identify volunteers as exceptional beings, and perhaps even elitist. Discussions surrounding one's privileged upbringing and their global awareness emerged in some of the interviews, and a sense of social responsibility and thoughtfulness was established.

As stated in the introduction, Wearing (2001) and Singh and Singh's (2004) popularly cited definitions of volunteer tourism neglect to examine 'what is in it?' for the traveller. Specifically, many of the existing definitions of volunteer tourism centralise the importance of the 'altruistic' characteristics of volunteers and neglect the hedonistic aspects that some travellers reflect. As such, they prioritise the outward focus of volunteering (e.g. Wearing 2001; Singh and Singh 2004; McGehee and Santos 2005) and neglect the inward focus that dominates the discourses explored in this study. Coincidently, one of the main findings in this study was the reduced concern and/or reflection the informants put forth with regard to the contribution they were directly making to the actual community. This challenges the part of Wearing's (2001: 1) definition that states that volunteer tourism might 'involve aiding or alleviating the material poverty of some groups in society'. Perhaps aiding or alleviating of poverty is carried out unconsciously by volunteer travellers?

The current definitions of volunteer tourism, then, resonate with the societal discourse that volunteering is good, virtuous and altruistic. In this study we have demonstrated that informants drew on this notion in their construction of a self-image as they employed discourse that enabled them to project an ethical self. In contrast, the analysis also demonstrated an inward focus on personal development and benefits. This study, then, illustrates the complexity of motivations to volunteer, and the process of volunteering.

Activity

Travellers often demonstrate a variety of motivations propelling them to make their travel-related decisions. Using specific discourses before, during and upon returning from one's holiday may further incentivise one's specific travel-related decisions. Such discourses may add stimulus to specific travel choices. Thus, travellers enacting their motivations, exploiting specific discourses, and being seen in a particular light may all be factors in the decision to participate in particular activities for a part of or all of their holiday.

Before class, ask everyone to produce a one-page account of a holiday they have had, or would like to have. In class, split the students into groups of three or four and ask them to analyse the discourses of each other's holiday stories, taking into account the following questions:

1. What type of holiday was described?
2. Which aspects of their holidays are the authors highlighting when recounting their holidays to others?
3. Are the authors willing to discuss which aspects of their holidays they would prefer to forget when recounting their stories?
4. What self-image are the authors projecting through their accounts?

Additional resources

www.cotlands.org.za/
www.africanimpact.com/responsible-travel/
www.bbc.co.uk/news/magazine-22294205

References

Andorfer, V.A. and Liebe, U. (2013) 'Consumer Behaviour in Moral Markets. On the Relevance of Identity, Justice Beliefs, Social Norms, Status and Trust in Ethical Consumption' in *European Sociological Review*, DOI:10.1093/esr/jct014.

Antaki, C., Billig, M., Edwards, D. and Potter, J. (2003) 'Discourse Analysis Means Doing Analysis: A Critique of Six Analytic Shortcomings' in *Discourse Analysis Online*, 1 (1). Online, available at: www.shu.ac.uk/daol/previous/v1/n1/index.htm.

Barbieri, C., Almeida Santos, C. and Katsube, Y. (2012) 'Volunteer Tourism: On-the-ground Observations from Rwanda' in *Tourism Management*, 33: 509–16.

Barnett, C., Cloke, P., Clarke, N. and Malpass, A. (2005) 'Consuming Ethics: Articulating the Subjects and Spaces of Ethical Consumption' in *Antipode*, 37 (1): 24–45.

Bennett, R. (2008) 'Booming Gap Year under Fire: Year-off Controversy. The Student guide' in *The Times*, 14 August 2008.

Billig, M. (1999) 'Commodity Fetishism and Repression: Reflections on Marx, Freud and the Psychology of Consumer Capitalism' in *Theory Psychology*, 9: 313–29.

Brown, S. (2005) 'Travelling with a Purpose: Understanding the Motives and Benefits of Volunteer Vacationers' in *Current Issues in Tourism*, 8 (6): 479–96.

Bryman, A. and Bell, E. (2007) *Business Research Methods*, 2nd edition, New York: Oxford University Press.

Bussell, H. and Forbes, D. (2002) 'Understanding the Volunteer Market: The What, Where, Who and Why of Volunteering' in *International Journal of Nonprofit and Voluntary Sector Marketing*, 7 (3): 244–57.

Callanan, M. and Thomas, S. (2004) 'Volunteer Tourism: Deconstructing Volunteer Activities Within a Dynamic Environment' in M. Novelli (ed.), *Niche Tourism: Contemporary Issues, Trends and Case Studies*, Amsterdam: Elsevier: 183–200.

Dhruvarajan, V. (2000) 'Globalisation: Discourse of Inevitability, and a Quest for a Just and Caring World' in *Society-Societe*, 24(2): 1–12.

Foucault, M. (1972) *The Archaeology of Knowledge and the Discourse on Language*, New York: Pantheon Books.

Foucault, M. (1987) 'The Ethic of Care for the Self as a Practice of Freedom: An Interview with Michel Foucault' in *Philosophy & Social Criticism*, 12 (1): 112–32.

Guttentag, D.A. (2009) 'The Possible Negative Impacts of Volunteer Tourism' in *International Journal of Tourism Research*, 11: 537–51.

Higgins-Desbiolles, F. (2006) 'More than an "Industry": The Forgotten Power of Tourism as a Social Force' in *Tourism Management*, 27. 1192–1208.

Holbrook, M.B. and Hirschman, E.C. (1982) 'The Experiential Aspects of Consumption: Consumer Fantasies, Feelings and Fun' in *Journal of Consumer Research*, 9 (2): 132–40.

McGehee, N.G. and Santos, C.A. (2005) 'Social Change, Discourse and Volunteer Tourism' in *Annals of Tourism Research*, 32 (3): 760–79.

McKinlay, A. and McVittie, C. (2008) *Social Psychology and Discourse*, Oxford: Wiley-Blackwell.

Morgan, J. (2009) 'Volunteer Tourism: What are the Benefits for International Development?' in *VolunTourist Newsletter*, 6(2). Online, available at: www.voluntourism.org/news-studyandresearch62.htm (accessed 14 August 2012).

Mustonen, P. (2005) 'Volunteer Tourism: Postmodern Pilgrimage' in *Journal of Tourism and Cultural Change*, 3 (3): 160–74.

Mustonen, P. (2007) 'Volunteer Tourism: Altruism or Mere Tourism?' in *Anatolia: An International Journal of Tourism and Hospitality Research*, 18 (1): 97–115.

Schaefer, A. and Crane, A. (2005) 'Addressing Sustainability and Consumption' in *Journal of Macromarketing*, 25 (1): 76–92.

Shaw, D. and Newholm, T. (2002) 'Voluntary Simplicity and the Ethics of Consumption' in *Psychology & Marketing*, 19 (2): 167–85.

Simpson, K. (2004) '"Doing Development": The Gap Year, Volunteer-tourists and a Popular Practice of Development' in *Journal of International Development*, 16 (5): 681–92.

Sin, H.L. (2009) 'Volunteer tourism "Involve Me and I Will Learn"' in *Annals of Tourism Research*, 36 (3): 480–501.

Singh, S. and Singh, T.V. (2004) 'Volunteer Tourism: New Pilgrimages to the Himalayas' in T.V. Singh (ed.), *New Horizons in Tourism: Strange Experiences and Stranger Practices*, Cambridge, MA: CAB International: 181–94.

Stebbins, R.A. (1992) *Amateurs, Professionals and Serious Leisure*, Montreal: McGill-Queens University Press.

Stoddart, H. and Rogerson, C.M. (2004) 'Volunteer Tourism: The Case of Habitat for Humanity South Africa' in *GeoJournal*, 60: 311–18.

Taylor, S. (2001) 'Locating and Conducting Discourse Analytic research' in M. Wetherell, S. Taylor and S. Yates (eds), *Discourse as data: A guide for analysis* London: Sage Publications: 5–48.

Varul, M.Z. (2009) 'Ethical Selving in a Cultural Context: Fair Trade Consumption as an Everyday Ethical Practice in the UK and Germany' in *International Journal of Consumer Studies*, 33: 183–89.

Wearing, S. (2001) *Volunteer Tourism: Experiences That Make a Difference*, Wallingford, Oxfordshire: CAB International.

Wearing, S. (2004) 'Examining Best Practice in Volunteer Tourism' in R. Stebbins and M. Graham (eds), *Volunteering as Leisure/Leisure as Volunteering: An International Assessment*, Wallingford, Oxfordshire: CAB International: 209–24.

Wetherell, M. and Potter, J. (1992) *Mapping the Language of Racism: Discourse and the Legitimation of Exploitation*, New York: Columbia University Press.

Wiggins, S. and Potter, J. (2008) 'Discursive Psychology' in C. Willig and W. Rogers (eds), *The Sage Handbook of Qualitative Research in Psychology*, London: Sage Publications: 73–90.

Zahra, A. and McIntosh, A. (2007) 'Volunteer Tourism: Evidence of Cathartic Tourist Experiences' in *Tourism Recreation Research*, 32 (1): 115–19.

10 Ethical tourism
The role of emotion

Sheila Malone

Introduction

The contradictions and inconsistencies of the tourism industry in relation to ethical consumption practices have never been more evident. Many authors have acknowledged the potentially negative impacts of a growing tourism industry (e.g. Fennell and Malloy 1995; Archer *et al.* 2005; Fennell 2006; Figueroa and Waitt 2011). These impacts are associated with cultural conflicts between host and guest (Smith and Brent 2001), environmental impacts related to landscapes (Hudson 2000), and more general concerns associated with overcrowding, water and air pollution, littering and waste overcapacity (Budeanu 2007). Such issues have attracted considerable attention, with ethics in tourism becoming a growing research area (Kalisch 2002; Holden 2003; Caruana and Crane 2011). Yet the place of ethics in tourism is somewhat fragmented. Many authors argue that the foundation of ethics-related tourism offerings is based on the concept of sustainable development (e.g. Weeden 2002; Butcher 2003) and, although an increase in demand for ethics-related tourism services is evident (Mintel 2011a, 2011b), it has not come without consequences.

The tourism industry has experienced immense growth in alternative, ethics-related tourism experiences, spanning a range of offerings such as ecotourism, responsible tourism and sustainable tourism. These offerings tend to reflect ethical efforts through practices such as eco-style accommodation, and less intrusive sports or transport as part of the holiday experience. However, Fennell (2006) calls for a more holistic approach to ethics-related tourism offerings, claiming that such practices are not only a process, but also form an ethic that takes into account all relevant stakeholders. Ongoing inquiries in the field of tourism are evident (e.g. Malloy and Fennell 1998a, 1998b; Weeden 2002; Miller and Hudson 2004; Miller and Twining-Ward 2005; Fennell 2006; Mowforth and Munt 2009; Pattullo and Minelli 2009) and in a recent study 72 per cent of respondents ranked choosing a holiday destination that preserves the culture and heritage of the host destination as very important (Mintel 2011a). However, despite these concerted efforts, a gap still exists with regard to our understanding of consumers' ethical choice processes, as they often act in contradiction to their expressed ethical concerns (Carrigan and Attalla 2001; Bergin-Seers and Mair 2009).

This chapter aims to address some of the issues outlined above by focusing on consumers' ethical behaviour in a tourism context. By offering an alternative view to the purely rational perspective of ethical choice, this chapter takes into account the role emotions play in tourists' ethical decision-making process. Lisle (2008) claims that alternative forms of tourism experiences are not a solution to the unethical actions associated with mainstream tourism practices, but they do enable an interrogation of consumer behaviour and perceived ethical activities. Although the existence of a more informed, aware, consumer is apparent (Mintel 2011b), our understanding of the factors that motivate or influence him/her is largely unknown. Thus, this chapter will demonstrate the importance of emotional experiences, not only in motivating and reinforcing tourists' ethical choices, but also as an integral part of the core consumption experience. Hence, the author offers an alternative viewpoint to the dominant rational perception of ethical choice and the purely ascetic interpretation of ethical tourism practices, thereby contributing to the literature on ethical tourism and ethical consumer behaviour theory.

Ethical tourism: an emotional choice

The Ethical Travel Guide, a publication by Tourism Concern, a not-for-profit UK charity, states that, by choosing one of the ethics-related hotels, experiences or treks identified in the guidebook, 'You can then have a great holiday and not take a guilt trip' (Pattullo and Minelli 2006: viii). This statement not only encapsulates the experiential nature of ethical tourism encounters but, more importantly, it illustrates the emotional conflict embedded in tourists' ethical decision-making process and ethical consumption practices. In general, tourism studies identify emotion as an evaluative construct associated with tourism destination loyalty or emotional satisfaction (e.g. Mano and Oliver 1993; Bigné and Andreu 2004; Wong 2004; Bigné *et al.* 2008), as a response to complaint experiences (Schoefer and Ennew 2005), as a segmentation tool based on the pleasure and arousal associated with leisure and tourism services, or as a motivator for holiday choice behaviour (Gnoth *et al.* 2000a, 2000b; Bigné and Andreu 2004; Bigné *et al.* 2005; del Bosque and San Martin 2008; see also Pike and Ryan's (2004) destination affective images and Hosany and Gilbert's (2010) destination emotion scale). Few studies have considered the influential or motivational role of emotion, in particular relating to tourism choices such as ethical, alternative or ecotourism.

The link between consumers' ethical attitudes and their behaviour has been explored by many academics. Employing Ajzen's Theory of Planned Behaviour (1991), Shaw and Clarke (1999) and Moons and De Pelsmacker (2012) suggest that a sense of ethical obligation helps individuals to choose appropriate behaviour based on what they feel they ought to do in terms of consumption choices or purchase behaviour. Rooted in deontological ethics, an individual's sense of responsibility emanates from internalised ethical beliefs and values, and an awareness of the implications of their actions for others.

However, 'ethical blindness' can occur, wherein an individual 'behave(s) unethically without being aware of it' (Palazzo et al. 2012: 109). Therefore, in order for an individual to feel ethically obliged to behave in a particular manner, a heightened level of ethical mindfulness, a commitment to a particular course of action, and an understanding of the impact of their actions are all required. Although not explored to date, it is conceivable that an individual's sense of ethical duty implies an emotional undertone, and that ethical behaviour is bound by emotional appeal due to the fact that these situations often lead to feelings of guilt (Marks and Mayo 1991; Pattullo and Minelli 2006), which, in turn, 'influence(s) the consumer's future (ethical) behaviour' (Marks and Mayo 1991: 721).

The Cognitive-Affective Model (Gaudine and Thorne 2001), founded on Rest's (1979) ethical decision-making framework, proposes that two dimensions of emotion are combined in individuals' ethical choice formation: emotional arousal and feeling state (valence). This model is based on the premise that 'ethical decisions often are emotionally charged; however, this does not necessarily suggest that the ethical decision process is not rational' (Gaudine and Thorne 2001: 175). The authors claim that high arousal and positive feeling states generate greater vigilance in situations where ethical dilemmas and reasoning are present. In situations with low arousal and positive feeling states, individuals are less likely to recognise the existence of an ethical issue, due to a lack of interest or motivation towards the task at hand. The authors maintain that positive emotions elicited during the decision-making process are often poor evaluators of an ethical dilemma, as individuals may want to maintain a positive emotional state, thus avoiding any action that may lead to a negative emotional outcome. Others (e.g. McCullough et al. 2001) also support the role of positive emotions, such as gratitude and pride, in influencing an individual's adherence to moral standards and pro-social behaviour.

The role of negative emotions in the ethical decision-making process of consumers has received much attention, but our understanding remains ambiguous (Marks and Mayo 1991; Gaudine and Thorne 2001; Connelly et al. 2004; Rajeev and Bhattacharyya 2007). Gaudine and Thorne (2001) suggest that negative emotions are complex due to the fact that some will signal the existence of an ethical dilemma (sadness), while others may not (depression). Psychologists, such as Baumeister et al. (2001), claim that negative emotions have the potential to motivate a more careful and thorough processing of information, leading to a more cautious consumption choice. However, studies to date tend to focus on individuals' experiences of negative emotions in the post-consumption stage, finding that feelings of guilt (Marks and Mayo 1991) or regret and disappointment (Rajeev and Bhattacharyya 2007) are experienced 'when behaviour and intentions are inconsistent with ethical judgments' (Ferrell et al. 1989: 60). A greater understanding of the role of emotion in consumers' ethical decision-making is imperative, due to its potential impact on future ethical consumption practices. However, to date, the findings regarding the meaning of positive and negative emotions are clearly mixed.

Emphasising a positive relationship between emotion and individuals' ethical intentions, Moons and De Pelsmacker (2012: 217) comment that 'emotional factors are important determinants of usage intention'. Due to the intention–behaviour gap evident in attitude-based models, given that expressed ethical concerns (intention) alone are insufficient to guarantee actual behaviour (Carrigan and Attalla 2001), the role and meaning of emotion in validating and reinforcing consumers' ethical choices are noteworthy. Despite the insight offered by Gaudine and Thorne's (2001) CAM, they claim that 'little is known about how emotions influence individuals' ethical decision making' (2001: 157). Therefore, taking into account Isen and Patrick's (1983) mood maintenance theory, and Frijda's (1986) action tendency, it can be argued that positive emotions can help validate and reinforce a particular course of action or approach behaviour (action tendency), as individuals aim for a positive emotional outcome (goal-directed behaviour: mood maintenance). Consequently, individuals' ethical choices are often authenticated by experiences of positive emotions. Similarly, negative emotions such as sadness, anger or guilt can have a positive effect on ethical choices as they may result in inaction or avoidance behaviour, or if a positive emotional outcome is desired, they may motivate behavioural change. For example, based on a need to maintain a positive emotional state, feelings of pride emanating from one's ethical behaviour (intended or actual) can influence and motivate ethical choices, as well as potentially affect or reinforce future ethical behaviour.

The emotional value embedded in ethical tourism is illustrated by Goodwin and Francis (2003: 273), who maintain that such experiences offer 'emotional recreation', motivated by a desire to 'feel good'. While tourists' experiences of ethical or responsible practices have attracted the attention of researchers, few efforts have focused on the emotive aspects of ethical choice, or its role in the consumption experience. Given that emotions can help resolve 'what they [individuals] do and what they say, between what they do and what seems most appropriate ... [and] between what they do and what they profess to know they should do' (Frijda 2008: 69), this area of investigation has an important contribution to make to the literature on ethical tourism and ethical consumer behaviour in terms of helping to understand the intention–behaviour gap identified in the literature by recognising the central role emotions play in tourists' ethical decision-making processes.

Ethical tourism: an enjoyable experience

The experiential perspective of consumer behaviour emphasises the hedonic and emotive aspects of consumption experiences and product usage (Hirschman and Holbrook 1982; Holbrook and Hirschman 1982). However, such consumption phenomena have been largely ignored, despite the fact that 'consumption experience is replete with emotion' (Elliott 1998: 96). Mowen (1988) maintains that consumers often make decisions in order to create feelings, experiences and emotions. Rooted in the experiential perspective of consumer behaviour, he argues that consumer choice is not necessarily a rational process aimed at problem solving per se, but is often an emotional experience motivated by a desire for pleasure.

It is plausible that the failings of ethical theorists and tourism scholars to date in predicting consumers' ethical behaviour may be attributed to the fact that they have overlooked non-rational factors such as emotion as a constituent part of the core consumption experience. The experiential and sensorial nature of tourism fosters an emotionally charged consumption experience (Pearce 2007). The idea that consumers' ethical choices are influenced by the emotive aspects of the core consumption experience is relatively novel. However, the visceral reactions of guilt and shame among tourists, as common emotional responses to travel and tourism experiences, and the tension tourism presents, attest to the emotions with which tourists' ethical consumption experiences are laden.

Ethical tourism is considered to be an act of compromise in the face of conflicting hedonic, social and environmental concerns (Fennell 2006). Butcher (2003: 7) suggests that the link between ethical tourism, fun and adventure is, more often than not, overshadowed by a new ethical imperative 'the association of tourism with innocence, fun and adventure, has been challenged by a mood of pessimism and a sense that moral regulation of pleasure-seeking is necessary in order to preserve environmental and cultural diversity'. Traditionally, mainstream tourism has been linked with hedonistic motivations (Gnoth *et al.* 2000a; Goossens 2000). Yet, its ethical counterpart is not usually associated with hedonic value. Goodwin and Francis (2003) claim that a shift in tourism consumption practices means consumers no longer desire the basic sun, sand and sea package, but want experiences, engagement, self-actualisation, fulfilment and rejuvenation, based on a desire to feel good. Thus, it can be argued that ethical alternatives are not an altruistic choice based on ascetic encounters per se, but that tourists choose ethical options in order to have pleasurable experiences and are motivated by hedonistic associations.

Buckley (2012: 535) conducted a review of the literature on sustainable tourism alluding to the long-standing paradoxical relationship between hedonism and ethical choice. He states that 'the future of tourism depends largely on conflicting social and economic pressures', and that 'people want holidays, and on holiday they act hedonistically'. However, to date, the most accepted drivers of consumers' ethical tourism choices tend to stem from an interest in people and the environment. It is thought that such decisions are driven by a personal desire for education, to offer communion, and/or so as to have cultural experiences (e.g. Goossens 2000; Molloy 2009), with little attention afforded to the notion that ethical tourism consumption may be motivated by emotive aspects of the consumption experience, defined as pleasure seeking.

By commenting that 'hedonism regards pleasure as the goal that renders participation in an activity worthwhile', Fennell (2006: 71) encapsulates the subjective nature of tourism experiences. That is, by 'allowing the individual to determine what is pleasurable' (ibid.), he demonstrates the difficulty of defining the role of pleasure in ethical tourism experiences. Theorists such as Szmigin and Carrigan (2006: 610) claim that 'ethical hedonism', that is, the pleasure derived from choosing an ethical alternative, often leads to 'feelings of pleasure from the purchase and in terms of the good they may bring to others'. This highlights the fact that hedonistic and rational motivations for ethical choices have become intertwined. Stemming from Gabriel and Lang's (1995) suggestion that consumers are creators of their own meaning, a new wave of hedonistic experiences is emerging. Furthermore, Soper (2007, 2008) and Soper et al. (2009: 572) maintain that ethical consumption can, in itself, be a hedonistic pursuit. Referring to the concept of 'alternative hedonism', they posit that consumers strive for the 'good life', defined as consuming in an alternative way that offers greater intrinsic satisfaction. Thus, pleasure is not only derived from choosing more ethically but is also found in the act of consumption.

The link between ethical tourism and hedonism has two credible bases. First, based on a desire to satisfy a consumer's ethical orientation through a deliberate 'pious and worthy' choice (Schaefer and Crane 2005), or an 'ethical hedonism' (Szmigin and Carrigan 2006: 610), feelings of pleasure stem from the decision-making process itself. Such choices complement and fulfil a consumer's desire to express their ethical beliefs and values. This is an active choice that is not only a mindset, but also a way of feeling. Second, pleasure is felt as part of the experience itself, through involvement and participation, termed an 'alternative hedonism' (Soper 2007, 2008; Soper et al. 2009). In this case, ethical tourism is an enriching experience and offers a more meaningful consumption encounter based on the fact that it is an intentional act that adds value in the form of the emotional benefits obtained as a result of exercising one's ethical beliefs and values. This is a noteworthy proposition, given that the role of hedonic value, defined as pleasure seeking, is largely unknown in ethical tourism, yet ethical choice appears to be steeped in emotional conflict (Marks and Mayo 1991; Gaudine and Thorne 2001; Pattullo and Minelli 2006) and ethical consumption experiences are replete with emotional value, often motivated by a desire for pleasure (Szmigin and Carrigan 2006; Soper 2007, 2008; Soper et al. 2009).

As 'our emotions more than reason, guarantee our commitment, and act as a basis for the moral choices we make' (Fennell 2006: 40), they have an important contribution to make to ethical decision-making theory, as they can pave the way for ethical cooperation. This area is ripe for investigation as Szmigin and Carrigan (2006: 612) explain when they say 'understanding and exploring the dimensions of consumer pleasure are a necessary prerequisite to furthering the development of ethical consumption'. On the whole, we know little about the emotive aspects of the ethical consumption experience. Nor do we know much about the impact of emotion on a tourist's ethical decision-making process, or its role in motivating particular tourism choices, such as ethical, alternative or ecotourism.

This is surprising given that tourism is a leisure activity, in which emotion often influences and motivates tourists' behaviour (e.g. Gnoth et al. 2000a, 2000b). The role of non-rational factors in motivating consumers' ethical choices has been overlooked, although socio-psychological explanations of contradictory behaviour have been posited (e.g. Chatzidakis *et al.* 2007). Many authors (e.g. Gaudine and Thorne 2001) support the need to include emotion in models of ethical choice, as a greater understanding of the role emotions play in ethical tourism practices will help further the development of ethical consumer behaviour theory by advancing our knowledge of consumers' ethical intentions and behaviour and thus offer an alternative to the limited ascetic view of ethical consumption.

Conclusion

This chapter has explored the role of emotion in the ethical decision-making process of tourists, and its presence in the consumption experience. In doing so, an alternative view to the dominant rational perspective of tourists' ethical choice formation, and the essentially ascetic interpretation of ethical tourism practices, has been proposed. The emergence of the ethical consumption concept is a growing area of concern for the tourism industry. This is magnified by the fact that there is much confusion with regard to what motivates consumers' ethical tourism choices, and what values are embedded in these experiences. It is further complicated by the fact that traditional theories of ethical decision-making appear to be deficient in terms of predicating ethical intention or behaviour. While ethical choice may seem to be a rational and deliberate act, this chapter has argued that the existing theories and decision-making models offered in the literature are inadequate as they overlook non-rational factors such as emotion.

As consumers' ethical choices are not bound by rational considerations alone, the author proposes that such decisions are often sealed by emotional engagement. By considering the emotive aspects of ethical tourism experiences, it is plausible that such choices are motivated by a need for pleasure and enjoyment. It can be concluded that ethical choice-making is frequently an emotion-driven process, in which tourists navigate and negotiate their way through a range of ethical considerations in order to express and successfully satisfy their ethical beliefs and values. It is in the process of negotiation that emotions may have the greatest role to play, as they can help motivate, validate and reinforce consumers' ethical choices. As consumers' ethical choice-making is authenticated by emotions, emotional experiences tend to influence their future ethical behaviour, leading them to avoid unethical actions and negative emotional outcomes. Nonetheless, numerous cognitive and emotional conflicts may occur, and concessions are often made as value systems collide with the needs of the marketplace.

This chapter contributes to the literature on ethical tourism and ethical consumer behaviour by exploring the concept of emotion in consumers' ethical choice formation, and its integral role in the consumption experience. As ethical consumption is enacted through marketplace engagement, the ethics of consumption should be a matter of priority for both tourism providers and governmental organisations if they wish to develop a more future-facing industry. From a practical point of view, a greater understanding of the role emotions play in consumers' ethical choice formation could assist ethical tourism providers in developing more customised ethical tourism offerings. For example, tourism providers might modify their promotional material, so as to communicate and reflect their ethical beliefs and values to existing consumers, as well as potential new customers. If emotions influence, motivate and reinforce tourists' ethical choices, it is plausible that, as a result of developing an emotional attachment to a particular tourism organisation, customer loyalty and commitment will ensue. Malär et al. (2011) highlight the significance of consumers' strong emotional attachment to products and services, stating that it enhances their sense of self, generating consumer commitment and loyalty as a result. Thus, by understanding how ethical tourism is experienced and the motivations behind such encounters, tourism providers should be able to offer more pleasurable consumer experiences by satisfactorily meeting their customers' needs, wants and desires. Greater customer retention may follow as consumers move from being merely interested in, to becoming advocates of, the ethical tourism organisation.

A ripe area for future investigation is to develop the concept of commitment in ethical consumption practice, and to examine individuals' ethical commitment to such practices and the impact on their lives in general. The concept of commitment implies a lasting orientation that leads to customer retention (Moorman et al. 1992; Morgan and Hunt 1994). It is often associated with a customer's desire to reject alternatives (Pritchard et al. 1999) and remain loyal to a particular brand or organisation. According to Meyer et al. (1993), individuals can have various reasons for committing: (1) because they want to feel emotionally attached (affective commitment), (2) because customers perceive no other options than to commit (continuance commitment), or (3) because they feel obliged to stay (normative commitment). Although Shaw and Clarke's (1999) notion of ethical obligation lends itself somewhat to the realm of normative commitment, Fullerton (2011) maintains that customers who feel a deeper sense of affective commitment generally demonstrate a heightened level of customer advocacy to a particular brand/company. Therefore, future investigations should explore the role of emotion in strengthening a person's commitment to ethical consumption practices, and the impact of consumer advocacy on ethical consumption behaviour in terms of developing strong customer relationships, trust and commitment (Lawer and Knox 2006) towards particular tourism organisations and destinations (Hosany and Gilbert 2010). Such studies would offer an interesting line of enquiry and could potentially lead to effective marketing communication campaigns for ethical tourism providers, and new marketing tactics for social marketing campaigns aimed at engendering change behaviour towards more ethical consumption practices.

Activity

Fennell (2006) claims that ethical tourism practices are not just a process but form an ethic that takes into account all relevant stakeholders; this begs the question: who is responsible for ethical tourism practices and initiatives?

In groups of five or six, prepare a ten-minute 'opinion' piece taking one of the following points of view:

1. Ethical tourism practices are the responsibility of tourists;
2. Ethical tourism services are the responsibility of tourism providers;
3. Ethical tourism initiatives are the responsibility of the government.

To help prepare your argument(s), take into consideration the following issues:

1. The role tourists, tourism providers and the government play in maintaining an ethical tourism industry.
2. Why do tourists' choose ethical tourism alternatives?
3. What are the motivational factors that influence tourists' ethical tourism choices?
4. Is ethical choice a purely rational altruistic choice, or is it sealed by emotions and a desire for pleasure?
5. Are all ethical tourism offerings an act of compromise, and a joyless consumption experience in the face of conflicting hedonistic, social and environmental concerns?
6. Is ethical tourism a product of individual choice, a marketing concept or an agenda set out by government initiatives?
7. Should tourism providers ensure that ethical initiatives are offered in tourism experiences?
8. How can governments encourage the creation and maintenance of an ethical tourism industry?

Suggested sources

www.tourismconcern.org.uk
www.ethicalconsumer.org
www.roughguide-betterworld.com
www.travelmatters.co.uk/ethical-tourism/
www.responsibletravel.com

References

Ajzen, I. (1991) 'The Theory Of Planned Behaviour' in *Organizational Behaviour and Human Decision Processes*, 50: 179–211.
Archer, B., Cooper, C. and Ruhanen, L. (2005) 'The Positive And Negative Impacts Of Tourism' in W.F. Theobald (ed.), *Global Tourism* (3rd edition), New York: Butterworth-Heinemann/Elsevier: 70–102.
Baumeister, R.F., Bratslavsky, E., Finkenauer, C. and Vohs, K.D. (2001) 'Bad Is Stronger Than Good' in *Review of General Psychology*, 5: 323–70.
Bergin-Seers, S. and Mair, J. (2009) 'Emerging Green Tourists In Australia: Their Behaviours And Attitudes' in *Tourism and Hospitality Research*, 9 (2): 109–19.
Bigné, E.J. and Andreu, L. (2004) 'Emotions In Segmentation: An Empirical Study' in *Annals of Tourism Research*, 31 (3): 682–96.

Bigné, E.J., Andreu, L. and Gnoth, J. (2005) 'The Theme Park Experience: An Analysis Of Pleasure, Arousal And Satisfaction' in *Tourism Management*, 26 (3): 833–45.

Bigné, E.J., Matilla, A.S. and Andreu, L. (2008) 'The Impact Of Experiential Consumption Cognitions And Emotions On Behavioural Intention' in *Journal of Service Marketing*, 22 (4): 303–25.

Buckley, R. (2012). 'Sustainable Tourism: Research And Reality' in *Annals of Tourism Research*, 39 (2): 528–46.

Budeanu, A. (2007) 'Sustainable Tourist Behaviour – A Discussion Of Opportunities For Change' in *International Journal of Consumer Studies*, 31: 499–508.

Butcher, J. (2003) *The Moralisation of Tourism: Sun, Sand ... and Saving the World?* London: Routledge.

Carrigan, M. and Attalla, A. (2001) 'The Myth Of The Ethical Consumer – Do Ethics Matter In Purchase Behaviour?' in *Journal of Consumer Marketing*, 18 (7): 560–77.

Caruana, R. and Crane, A. (2011) 'Getting Away From It All' in *Annals of Tourism Research*, 38 (4): 1495–1515.

Chatzidakis, A., Hibbert, S. and Smith, A. (2007) 'Why People Don't Take Their Concerns About Fair Trade To The Supermarket: The Role Of Neutralisation' in *Journal of Business Ethics*, 74 (1): 89–100.

Connelly, S., Helton-Fauth, W. and Mumford, D.M. (2004) 'A Managerial In-Basket Study Of The Impact Of Trait Emotions On Ethical Choice' in *Journal of Business Ethics*, 51: 245–67.

del Bosque, I.R. and San Martin, H. (2008) 'Tourist Satisfaction. A Cognitive-Affective Model' in *Annals of Tourism Research*, 35 (2): 551–73.

Elliott, R. (1998). 'A Model Of Emotion-Driven Choice' in *Journal of Marketing Management*, 14: 95–108.

Fennell, D.A. (2006) *Tourism Ethics*, Clevedon: Channel View Publications.

Fennell, D. and Malloy, D.C. (1995) 'Ethics And Ecotourism: A Comprehensive Ethical Model' in *Journal of Applied Recreation Research*, 20 (3): 63–183.

Ferrell, O.C., Gresham, L.G. and Fraedrich, J. (1989) 'A Synthesis Of Ethical Decision Models For Marketing' in *Journal of Macromarketing*, 11: 55–64.

Figueroa, R.M. and Waitt, G. (2011) 'The Moral Terrains Of Ecotourism And The Ethics Of Consumption' in T. Lewis and E. Potter (eds.), *Ethical Consumption: A Critical Introduction*, London: Routledge: 260–75.

Frijda, N.H. (1986) *The Emotions*, London: Cambridge University Press.

Frijda, N.H. (2008) 'The Psychologists' Point Of View' in M. Lewis, J.M. Haviland-Jones, and L.F. Barrett (eds.), *Handbook of Emotions* (3rd edition), London: The Guildford Press: 66–88.

Fullerton, G. (2011) 'Creating Advocates: The Roles Of Satisfaction, Trust And Commitment' in *Journal of Retailing and Consumer Services*, 18: 92–100.

Gabriel, Y. and Lang, T. (1995) *The Unmanageable Consumer*, London: Sage Publications.

Gaudine, A. and Thorne, L. (2001) 'Emotion And Ethical Decision-Making In Organisations' in *Journal of Business Ethics*, 31 (2): 175–87.

Gnoth, J., Zins, A., Lengmueller, R. and Boshoff, C. (2000a) 'The Relationship Between Emotions, Mood And Motivation To Travel: Towards A Cross-Cultural Measurement Of Flow' in A.G. Woodside, G.I. Crouch, J.A. Mazanec, M. Oppermann and M.Y. Sakai (eds), *Consumer Psychology of Tourism Hospitality and Leisure*, Wallingford, Oxfordshire: CAB International: 155–75.

Gnoth, J., Zins, A., Lengmüller, R. and Boshoff, C. (2000b) 'Emotions, Mood, Flow And The Motivation To Travel' in *Journal of Travel and Tourism Marketing*, 9 (3): 23–34.

Goodwin, H. and Francis, J. (2003) 'Ethical And Responsible Tourism: Consumer Trends In The UK' in *Journal of Vacation Marketing*, 9 (3): 271–82.

Goossens, C. (2000) 'Tourism Information And Pleasure Motivation' in *Annals of Tourism Research*, 27 (2): 301–21.

Hirschman, C.E. and Holbrook, B.M. (1982) 'Hedonic Consumption: Emerging Concepts, Methods And Propositions' in *Journal of Marketing*, 46: 92–101.

Holbrook, B.M. and Hirschman, C.E. (1982) 'The Experiential Aspect Of Consumption: Consumer Fantasies, Feelings And Fun' in *Journal of Consumer Research*, 9: 132–40.

Holden, A. (2003) 'In Need Of A New Environmental Ethic For Tourism' in *Annals of Tourism Research*, 30 (1): 94–108.

Hosany, S. and Gilbert, D. (2010) 'Measuring Tourists' Emotional Experiences Toward Hedonic Holiday Destinations' in *Journal of Travel Research*, 49 (4): 513–26.

Hudson, S. (2000) *Snow Business: A Study of the International Ski Industry*, London: The Continuum International Publishing Group.

Isen, A.M. and Patrick, R. (1983) 'The Effect Of Positive Feelings On Risk Taking: When The Chips Are Down' in *Organizational Behaviour and Human Performance*, 31 (2): 194–202.

Kalisch, A. (2002) *Corporate Futures: Social Responsibility in the Tourism Industry*, London: Tourism Concern.

Lawer, C. and Knox, C. (2006) 'Customer Advocacy And Brand Development' in *Journal of Product and Brand Management*, 15 (2): 121–9.

Lisle, D. (2008) *Joyless Cosmopolitans: The Moral Economy of Ethical Tourism*, ISA's 49th Annual Convention, Building Multiple Divides, San Francisco, USA. Online, available at: www.allacademic.com/one/isa/isa08/index.php?click_key=1&PHPSESSID=6280365d352766b07401ecf3a33451a2 (accessed 16 March 2010).

Malär, L., Krohmer, H., Hoyer, W.D. and Nyffenegger, B. (2011) 'Emotional Brand Attachment And Brand Personality: The Relative Importance Of The Actual And The Ideal Self' in *Journal of Marketing*, 75 (4): 35–52.

Malloy, C.D. (2009) 'Can One Be An Unethical Ecotourist? A Response To R. Buckley's "In Search Of The Narwhal"' in *Journal of Ecotourism*, 8 (1): 70–3.

Malloy, C.D. and Fennell, D.A (1998a) 'Codes Of Ethics And Tourism: An Exploratory Content Analysis' in *Tourism Management*, 19 (5): 453–61.

Malloy, C.D. and Fennell, D.A. (1998b) 'Ecotourism And Ethics: Moral Development And Organisational Cultures' in *Journal of Travel Research*, 36: 47–56.

Mano, H. and Oliver, L. R. (1993) 'Assessing The Dimensionality And Structure Of The Consumption Experience: Evaluation, Feeling, And Satisfaction', *Journal of Consumer Research* 20: 451–66.

Marks, J.L. and Mayo, M. (1991) 'An Empirical Test Of A Model Of Consumer Ethical Dilemmas' in *Advances in Consumer Research*, 18: 720–7.

McCullough, M.E., Kilpatrick, S.D., Emmons, R.A. and Larson, D.B. (2001). 'Is Gratitude A Moral Affect?' in *Psychological Bulletin*, 127: 249–66.

Meyer, J.P., Allen, N.J. and Smith, C.A. (1993) 'Commitment To Organizations And Occupations: Extension And Test Of A Three-Component Conceptualization' in *Journal of Applied Psychology*, 78 (4): 538–51.

Miller, G. and Husdon, S. (2004) 'Ethical Considerations In Sustainable Tourism' in W. Theobald (ed.), *Global Tourism: The Next Decade*, Oxford: Butterworth-Heinemann: 248–60.

Miller, G. and Twining-Ward, L. (2005) *Monitoring for a Sustainable Tourism Transition: The Challenge of Developing and Using Indicators*, Wallingford, Oxfordshire: CAB International.

Mintel (2011a) *Eco-Accommodation in Europe*. Online, available at: www.mintel.com (accessed on 29 January 2012).

Mintel (2011b) *Green Tourism Innovations*. Online, available at: www.mintel.com (accessed on 29 January 2012).

Moons, I. and De Pelsmacker, P. (2012) 'Emotions As Determinants Of Electric Car Usage Intention' in *Journal of Marketing Management*, 28 (3–4): 195–237.

Moorman, C., Deshpande, R. and Zaltman, G. (1992) 'Relationships Between Providers And Users Of Market Research: The Dynamics Of Trust Within And Between Organisations' in *Journal of Marketing Research*, 29: 314–28.

Morgan, R.M. and Hunt, S. (1994) 'The Commitment-Trust Theory Of Relationship Marketing' in *Journal of Marketing*, 58 (3): 20–38.

Mowen, C.J. (1988) 'Beyond Consumer Decision Making' in *The Journal of Consumer Marketing*, 5 (1): 15–25.

Mowforth, M. and Munt, I. (2009) *Tourism and Sustainability: Development and New Tourism in the Third World* (2nd edition), London: Routledge.

Palazzo, G., Krings, F. and Hoffrage, U. (2012) 'Ethical Blindness' in *Journal of Business Ethics*, 109 (3): 323–38.

Pattullo, P. and Minelli, O. (2006) *The Ethical Travel Guide*, London: Earthscan/Tourism Concern.

Pattullo, P. and Minelli, O. (2009) *The Ethical Travel Guide* (2nd edition), London: Earthscan/Tourism Concern.

Pearce, P.L. (2007) *Tourist Behaviour: Themes and Conceptual Schemes*, Clevedon: Channel View Publications.

Pike, S. and Ryan, C. (2004) 'Destination Positioning Analysis Through A Comparison Of Cognitive, Affective And Conative Perceptions' in *Journal of Travel Research*, 42 (4): 333–42.

Pritchard, M., Havitz, M. and Howard, D. (1999) 'Analyzing The Commitment-Loyalty Link In Service Relationships' in *Journal of the Academy of Marketing Science*, 27 (3): 333–48.

Rajeev, P. and Bhattacharyya, S. (2007) 'Regret And Disappointment: A Conceptualisation Of Their Role In Ethical Decision-Making' in *Vikalpa*, 32 (4): 75–85.

Rest, J.R. (1979) *Development in judging moral issues*, Minneapolis, MI: University of Minnesota Press.

Schaefer, A. and Crane, A. (2005) 'Addressing Sustainability And Consumption' in *Journal of Macromarketing*, 25 (1): 76–92.

Schoefer, K. and Ennew, C. (2005) 'The Impact Of Perceived Justice On Consumers' Emotional Responses To Service Complaint Experiences' in *Journal of Services Marketing*, 19 (5): 261–70.

Shaw, D. and Clarke, I. (1999) 'Belief Formation In Ethical Consumer Groups: An Exploratory Study' in *Marketing Intelligence & Planning*, 17 (2): 109–19.

Smith, V.L. and Brent, M. (2001) *Hosts and Guests Revisited: Tourism Issues of the 21st Century*, New York: Cognizant.

Soper, K. (2007) 'Rethinking The "Good Life"' in *Journal of Consumer Culture*, 7 (2): 205–29.

Soper, K. (2008) '"Alternative hedonism" and the citizen consumer' in K. Soper and F. Trentmann (eds.), *Citizens and Consumption*, Hampshire: Palgrave Macmillan: 191–234.

Soper, K., Ryle, M. and Lyn, T. (2009) *The Pleasures and Politics of Consuming Differently: Better Than Shopping*, London: Palgrave Macmillan.

Szmigin, I. and Carrigan, M. (2006). 'Exploring The Dimensions Of Ethical Consumption' in *European Advances in Consumer Research*, 7: 608–12.

Weeden, C. (2002) 'Ethical Tourism: An Opportunity For Competitive Advantage?' in *Journal of Vacation Marketing*, 8 (2): 141–53.

Wong, A. (2004) 'The Role Of Emotional Satisfaction In Service Encounters' in *Managing Service Quality*, 14 (5): 365–76.

Part III
Helping consumers make ethical decisions

11 Tread lightly through this *Lonely Planet*

Examining ethical information in travel guidebooks

Sarah Quinlan Cutler

Introduction

Guidebooks are a much-used and influential information source for many tourists. These texts present information which allows a visitor to physically, economically, culturally and environmentally negotiate a destination, yet the examination of guidebooks and the role they play in tourism has received little attention in the academic literature. The purpose of this chapter is to discuss how travel guidebooks present and communicate information on ethical tourism. Three commonly used English language commercial guidebooks on Peru are evaluated with regard to the content of ethical information provided. This chapter stresses the importance of examining these texts and the possible role they play in promoting ethical tourism.

Tourist information sources

Tourists face a lot of options when it comes to deciding where to go and what to do during a tourism event, and many rely on outside sources, such as web sites, brochures and travel guidebooks, for information that will ultimately influence how they spend their time and money. The search for information is an important component of tourist decision-making (Wong and Liu 2011). Tourism information sources inform tourists about services provided at destinations such as accommodation, transportation, activities, and food and beverages. Tourism information sources can also act as educative tools, improving tourist understanding of a destination by providing information on culture, religion, environment, history and local attractions. A better understanding of destinations can enhance the quality of travel experiences, which is why prospective tourists devote time and attention to searching out destination information (Osti *et al.* 2009).

As a tourist information source, the travel guidebook has been part of tourism ever since people began traveling for leisure (Otness 1993). What would be considered a modern guidebook has been employed by travelers since the nineteenth century, when guidebooks began to provide maps, advice, details of attractions and suggestions for itineraries (Gilbert 1999). Guidebooks are argued to be significant elements of the tourism infrastructure, influencing the perception of destinations and the travel practices of millions of tourists (Kosher 1998; Gilbert 1999). Lew (1991) argues that at a destination, guidebooks provide interpretation through the presentation of spatial and social information and by emphasizing attractions and destination characteristics. By stressing certain places or objects, guidebooks define both desirable and undesirable aspects and experiences (Lew 1991; Carter 1998). However, there have been few studies on the role of travel guidebooks in tourism (Tegelberg 2010).

The existing research that specifically looks at guidebooks identifies these texts as influential (Bhattacharyya 1997; McGregor 2000; Tegelberg 2010). McGregor (2000) argues that guidebooks open up foreign destinations by providing information on their accessibility and attractiveness while also structuring the way we travel. These texts construct perceptions of places for travelers before they have traveled, providing an underlying ideology of destinations (Carter 1998). Guidebook text is argued to provide a lens for viewing the world (McGregor 2000). Bhattacharyya (1997) analyzed the narratives in *Lonely Planet India*, focusing on how local culture is represented. The author argues that guidebooks are a type of culture broker or mediator, negotiating the relationship between tourist and destination and between tourist and host population. This guidebook role encourages self-sufficiency in the tourist, thereby reducing their need to interact with or depend on local people while at the destination (Bhattacharyya 1997).

In Tegelberg's (2010) analysis of the Lonely Planet guide to Cambodia, the author goes further, concluding that the guidebook not only reduces the need for interaction with local people, it silences local perspectives and avoids contentious local issues. Instead the guidebook favors less controversial perceptions of the destination, and local culture, as a way of appealing to tourists. In Iaquinto's (2011) study, from the perspective of guidebook writers, the author notes that guidebooks are purposely produced by editors to appeal more to mainstream tourism audiences. Iaquinto goes on to argue that a physical guidebook is simply unable to include the large range of diverse perspectives that can be found in other new media. Through this inclusion and omission of information and perspectives, guidebooks exert a massive influence on the tourist experience (McGregor 2000). Bhattacharyya (1997) argues that through the use of the guidebook's narrative voice, these texts claim authority. This authority can be related to directing tourists within a destination, but could also be used to guide tourism practices at a destination.

One of the concerns noted by some researchers is that though tourists are often cited as a major cause of negative tourism impacts, initiatives aimed at improving tourist behavior through awareness are rarely implemented (Krippendorf 1987; Moscardo 1996). In reviews of tourist information sources there is little ethical tourism information provided. Santos (2004) found that tourism information in newspapers mostly described attractions and vacation options, with very limited ethical information. Pennington-Gray and Thapa (2004) analyzed the content of 264 Destination Management Organization (DMO) websites and found that few of them are educating tourists about cultural awareness and cultural responsibility. In Tegelberg's (2010) analysis of the Lonely Planet guide to Cambodia, the author found that though the publication claimed to promote responsible travel, it was limited in terms of social awareness. Therefore, questions remain as to what ethical tourism information is available and how this ethical information is conveyed to tourists. The United Nations World Tourism Organization's (UNWTO) *Global Code of Ethics for Tourism* (2001) provides a framework for ethical tourism development. Based on this document, the UNWTO outlines eight principles, which act as a guide for the tourist (UNWTO n.d.). These principles focus on social, environmental and economic considerations as well as other practical guidelines to help individuals have a more rewarding tourist experience. Ethical tourism involves the minimizing of negative tourism impacts and the maximizing of positive tourism impacts. It also focuses on providing enjoyable experiences for the tourist through "more meaningful connections with local people, and a greater understanding of local cultural and environmental issues" (ICRT Canada 2001). Growth in ethical and Fair Trade marketplaces suggests that consumer attitudes are changing towards consumption of more socially and environmentally responsible products, services and/or experiences (Goodwin and Francis 2003). This would indicate that there is a demand for ethical tourism initiatives. The purpose of this paper is to examine the role of travel guidebooks in communicating ethical tourism information.

Research methods

This research examines published text in commercial guidebooks on Peru, focusing on ethical content. An international visitor survey listed the three most popular guidebooks used as information sources by tourists visiting Peru (PromPerú 2003). In 2004 the updated versions of these publications were: *Lonely Planet: Peru* (Rachowiecki and Beech 2004), *South American Handbook* (Box 2004) and *The Rough Guide to Peru* (Jenkins 2003). A thorough content analysis of each of these publications was undertaken. Content analysis is deemed to be one of the most common approaches in geographical research on travel literature, allowing for the understanding of textual representations (Lew 1991). This method is used for measuring, classifying and evaluating the content of human communication (Mehmetoglu and Dann 2003). Analysis of content can include manifest content (which is the visible wording and/or explicit themes) and latent content (which is the underlying meaning in communication) (Babbie 2001). An advantage of this technique is that it is an unobtrusive assessment of image or text where the presence of the researcher does not influence the subject under study (Babbie 2001). However, content analysis is limited to the examination of communication records (Babbie 2001) and this data cannot be subject to further experimentation and therefore inferences related to causality are highly subjective (Mehmetoglu and Dann 2003).

In 2004 the three guidebooks were read in their entirety and data were broken down into statements, which were defined as full sentences or groups of consecutive sentences which addressed the same issue. An open coding process was used to identify ethical tourism content. This content was then re-examined through an axial coding process which grouped ethical statements into categories based on emergent subjects. Along with categorization of subjects, statements were also coded by theme and statement type. To qualify themes, statements were examined as to whether they discussed environmental and ecological issues, socio-cultural issues, or economic issues. Statement type was determined by evaluating the statement wording using the coding scheme outlined in Table 11.1. These descriptions of statement types are based on Dann's (2003) research on the characteristics of tourist signs or notices. Dann's study evaluated the wording qualities of signs and notices to create a typology providing categories based on increasing levels of social control. In this paper these categories have been revised and include only those which are applicable to travel information found in guidebooks (see Table 11.1).

The researcher was the sole coder to limit variability and to maintain mutually exclusive categorizations (see Lew 1991). A successful intercoder reliability test was performed by additional coders which found an acceptable level of

Table 11.1 Coding scheme for statement-type categorization

Statement type	Description	Example
Descriptive	Informative, descriptive or observational statements, providing details and educational information.	"The lodge is owned by the Ese'eja Indians of Sonene who provide guiding and cultural services." (*Lonely Planet: Peru*, p. 366)
Optional	Optional or demonstrative statements providing choices, opportunities or outlining local norms and/or routines.	"Similarly, travellers can promote patronage and protection of important archaeological sites and heritage through their interest and contributions via entrance and performance fees." (*South American Handbook*, p. 49)
Advisory	Advisory or participatory statements such as advice, suggestions, tips or polite instruction which endorse a specific activity or action.	"The churches often are open for early morning mass, as well as evening masses, but visitors at those times should respect the worshippers and not wander around." (*Lonely Planet: Peru*, p. 262)
Obligatory	Statements outlining laws, rules or warnings which can be threaten, forbid or encourage a particular course of action.	"Note too that you should always use a camping stove – campfires are strictly prohibited in Huascaran National Park, and wood is scarce anyway." (*Rough Guide to Peru*, p. 305)

Source: Adapted from Dann (2003).

agreement between coders (≥0.70 or 70 percent), indicating that multiple individuals can have similar results using the same coding scheme.

In 2012 the newest editions of these publications (see Box 2011; Miranda *et al.* 2011; Jenkins 2012) were again analyzed to assess updated information and changes in guidebook content. Similarities and differences found between the 2012 and 2004 publications are outlined by publication, and the implications of the changes are considered in the Discussion section.

Results

2004 guidebook publications

The newest guidebooks versions, as of 2004, were analyzed for ethical content, which was recorded and coded by subject matter. Each statement was also coded by publication source, book section, statement theme and statement type. Table 11.2 provides a summary of these four coding categories. The *Lonely Planet: Peru* (henceforth *Lonely Planet*) and *Rough Guide to Peru* (henceforth *Rough Guide*) had virtually the same number of statements on ethical tourism (n=177, n=178), while the *South American Handbook* contained fewer statements (n=119). The majority of statements were located in regional sections (60.3 percent), and the fewest in concluding sections (11.6 percent). Almost half of the ethical statements were based on socio-cultural themes (47.3 percent), with 35.4 percent relating to environmental themes, and 17.3 percent relating to economic themes. In evaluating statement type, 11.6 percent of ethical statements were obligatory, stating rules, warnings or laws; 25.3 percent were advisory, offering friendly tips or suggestions; 17.1 percent provided optional information statements, presenting choices or demonstrative explanations; and 46 percent were descriptive, providing details and educational information on various ethical themes.

Table 11.2 Summary of ethical guidebook coding categories

Variable	N	%	Variable	N	%
Source			*Theme*		
Lonely Planet	177	37.3	Environmental	168	35.4
SA Handbook	119	25.1	Socio-cultural	224	47.3
Rough Guide	178	37.6	Economic	82	17.3
Total	474	100.0	Total	474	100.0
Section			*Statement Type*		
Introduction	133	28.1	Descriptive	218	46.0
Regional sections	286	60.3	Optional	81	17.1
Concluding sections	55	11.6	Advisory	120	25.3
			Obligatory	55	11.6
Total	474	100.0	Total	474	100.0

All guidebook paragraphs were assigned a number and the paragraph number for each ethical statement was also recorded. Some 357 paragraphs were identified as having ethical statements from a total of 6,258 paragraphs for all three publications, indicating that on average 5.7 percent of guidebook content contained information related to ethical tourism.

Ethical statement themes

A Pearson's chi square test was conducted to evaluate the frequency distribution of ethical themes in guidebook publications. The test is based on the expectation of even distributions, and compares expected counts with observed counts. Results showed that the *Lonely Planet* and *Rough Guide* were comparable in their theme distribution (see Figure 11.1), though the *Rough Guide* did have fewer economic-themed statements (11.8 percent, n=21) than expected (Pearson Chi square = 12.800, p<0.05). The main difference lies with the *South American Handbook* which had fewer environmental-themed statements than expected (26.1 percent, n=31) and more economic statements (26.1 percent, n=31) when compared to the other publications.

The ethical theme distribution in each guidebook section was also evaluated. When evaluating the statements by section, it was found that the introduction had fewer environmental statements (24.8 percent, n=33) and more socio-cultural statements (62.4 percent, n=83) than expected (Pearson Chi square = 31.254, p<0.001), indicating that the majority of ethical statements in introductory sections were based on socio-cultural themes. Regional

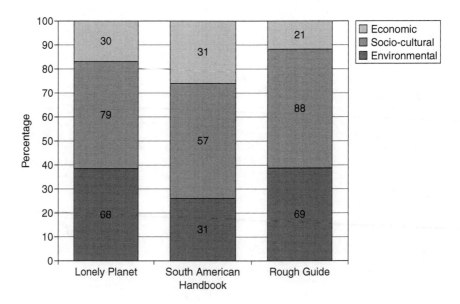

Figure 11.1 Themes of ethical statements by guidebook.

sections had more environmental statements (41.6 percent, n=119) and more economic statements (21.3 percent, n=61) but fewer socio-cultural statements (37.1 percent, n=106) than expected. Concluding sections mimicked introduction sections, having mostly socio-cultural-themed statements (63.6 percent, n=35).

Ethical statement type

Statements were coded based on four statement type categories (see Table 11.1 for coding scheme). Overall, 36.9 percent (n=175) of ethical statements provided persuasive messages (11.6 percent obligatory, 25.3 percent advisory) directly encouraging or commanding the tourist to make a particular choice or undertake a particular action (see Table 11.1 for details). Some 17.1 percent were optional statements, providing choices to tourists, though the wording tended to favor one choice over the other, indicating an indirect attempt at persuasion. The remainder of the statements (46.0 percent) were coded as descriptive, having no influential or persuasive quality (see Table 11.1 for details).

A Pearson chi square test was conducted to determine whether there were any statistically significant differences between expected and observed distributions of themed statements and statement types (see Figure 11.2). Results indicated that environmental-themed information had more obligatory (23.2 percent, n=39) and descriptive statements (57.1 percent, n=96) than expected, and fewer advisory (12.5 percent, n=21) and optional statements (7.1 percent) n=12) than expected (Pearson chi square = 74.998, p<0.001) indicating that

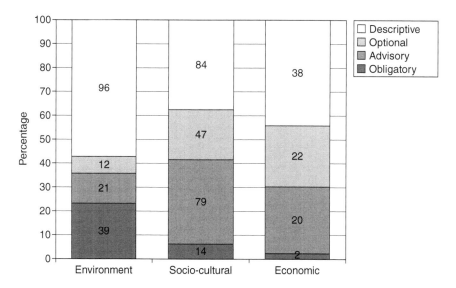

Figure 11.2 Themed statements showing ethical statement types.

environmental ethical statements tended to have either a very strong persuasive element or none at all. Socio-cultural-themed information had more advisory (35.3 percent, n=79) and fewer obligatory statements (6.3 percent, n=14) than expected.

Statement types were also evaluated for each guidebook source (see Figure 11.3). Again significant differences between observed and expected counts were found (Pearson chi square = 52.214, p<0.001). The *Lonely Planet* contained more obligatory (16.9 percent, n=30) and advisory statements (28.8 percent, n=51) than expected, but fewer descriptive statements (38.4 percent, n=68). The *South American Handbook* had more advisory statements (38.7 percent, n=46) but fewer descriptive statements (29.4 percent, n=35) than expected. The *Rough Guide* had more descriptive statements (64.6 percent, n=115) and fewer obligatory (6.7 percent, n=12) and advisory statements (12.9 percent, n=23) than expected. In evaluating each statement type specifically, more than half of all obligatory statements were found in the *Lonely Planet* guide (54.5 percent) whereas advisory statements were spread between the *Lonely Planet* (42.5 percent) and the *South American Handbook* (30.9 percent), with 19.2 percent found in the *Rough Guide*. Optional statements were more-or-less equal between all three guidebooks, but the majority of descriptive statements (52.8 percent) were located in the *Rough Guide*. This indicates that the *Rough Guide* was the text with the least persuasive language, providing tourists with ethical information rather than advice or mandatory action regarding ethical decision-making.

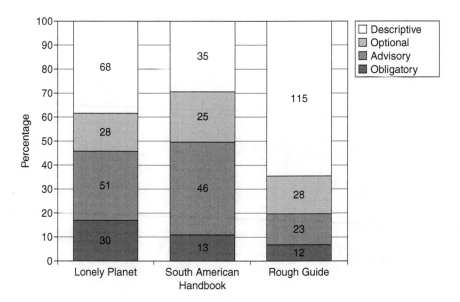

Figure 11.3 Guidebooks by ethical statement type.

Ethical content categories

The content analysis of the guidebooks resulted in 16 emergent ethical subject categories. These categories are listed in Table 11.3. The categories covered environmental, socio-cultural and economic themes regarding ethical tourism.

A cross-tabulation was completed to better understand where these categories were found in guidebooks. Tourist and local people interactions, contemporary culture, tourist temperament, impacts of tourism, and trade in flora, fauna and artifacts were more heavily weighted in the introduction sections whereas socio-cultural issues were more heavily weighted in conclusion sections.

When the spread of subject categories was compared in each publication, it was found that the *Lonely Planet* contained the majority of information on wildlife, contemporary culture, fire and language, whereas the *South American Handbook* contained the majority of statements on tourist temperament, impacts of tourism, and volunteering and donations, with very few statements on contemporary culture or wildlife. The *Rough Guide* contained the majority of statements on conservation and environmental issues and socio-cultural issues.

Statement types in each category were also examined. Seven subject categories were identified as having the majority of statements being persuasive in nature (obligatory and/or advisory statements). These categories were drugs and legal matters (100.0 percent), fire (100.0 percent), tourist temperament (78.5 percent), litter and waste (72.8 percent), trade in flora, fauna and artifacts (62.5 percent), and vandalism (57.2 percent). Volunteering and donations was found to contain a high percentage of optional statements (66.7 percent), and six categories were found to contain a high percentage of descriptive statements. These categories were socio-cultural issues (84.2 percent), conservation and environmental issues (73.2 percent), contemporary culture (69.2 percent), impacts of tourism (58.3 percent), language (55.6 percent), and wildlife (53.6 percent).

To determine how well the ethical subject categories fit within an ethical tourism framework, the categories were compared to the Responsible Tourist Principles outlined by UNWTO (see UNWTO, n.d.). Table 11.4 identifies each principle and lists the corresponding subject category found in guidebooks. Overall, many of the content categories adequately represent one or more principles of responsible tourism. The impacts of tourism category had overarching information on responsible tourism and could be applicable to all categories so it was left out of Table 11.4. None of the guidebook categories directly related to respecting human rights or sexual exploitation. The sixth principle regarding health did not appear in the emergent content categories, which was mainly due to the coding process. Each guidebook did have a specific section addressing health concerns, but this information was not coded in this research, which focused on environmental, socio-cultural and economic information.

Table 11.3 Frequency and valid percentage for subject categories identified in ethical guidebook statements

Subject categories	Category description	N	%
Conservation and environmental issues	Statements discussing the natural environment, sustainability, environmental management, environmental awareness or protected areas	97	20.5
Tourist and local people interactions	Statements discussing cultural practices and exchanges, cultural respect, photography of local peoples, and the treatment of porters.	90	19.0
Economic support	Statements discussing financial support for local economies, spending, budgeting, wealth and poverty, or bargaining	73	15.4
Society and cultural issues	Statements discussing overarching social or cultural issues or indigenous issues occurring at the destination	38	8.0
Wildlife	Statements discussing undomesticated fauna, hunting, fishing, expectations of wildlife, or wildlife disturbance	28	5.9
Language	Statements discussing language or language education	27	5.7
Contemporary culture	Statements discussing the current cultural climate, cultural differences, cultural similarities and social realities	26	5.5
Litter and waste	Statements discussing garbage, littering or waste management	22	4.6
Trade in flora, fauna and artifacts	Statements discussing trafficking and/or trading of natural or cultural items	16	3.4
Tourist temperament	Statements which address favorable or unfavorable tourist attitudes or disposition	14	3.0
Impacts of tourism	Statements discussing impacts of individual tourists or of the tourism industry	12	2.5
Volunteering and donations	Statements discussing volunteering or donations	12	2.5
Vandalism	Statements discussing physical damage, graffiti or defacing property	7	1.5
Fire	Statements discussing fire or fire regulations	6	1.3
Complaints	Statements discussing demands or complaints related to services and/or products in tourism	4	0.8
Drugs and legal matters	Statements discussing drug use and legal issues	2	0.4
Total		474	100

Table 11.4 Relating guidebook ethical subject categories to the principles of responsible tourism

UNWTO principle	Corresponding content categories
1 Open your mind to other cultures and traditions – it will transform your experience, you will earn respect and be more readily welcomed by local people. Be tolerant and respect diversity – observe social and cultural traditions and practices.	Tourist and local people interactions, Contemporary culture, Tourist temperament, Language.
2 Respect human rights. Exploitation in any form conflicts with the fundamental aims of tourism. The sexual exploitation of children is a crime punishable in the destination or at the offender's home country.	
3 Help preserve natural environments. Protect wildlife and habitats and do not purchase products made from endangered plants or animals.	Conservation and environmental issues, Wildlife, Litter and waste, Trade in flora, fauna and artifacts.
4 Respect cultural resources. Activities should be conducted with respect for the artistic, archaeological and cultural heritage.	Society and cultural issues, Trade in flora, fauna and artifacts.
5 Your trip can contribute to economic and social development. Purchase local handicrafts and products to support the local economy using the principles of fair trade. Bargaining for goods should reflect an understanding of a fair wage.	Economic support, Volunteering and donations.
6 Inform yourself about the destination's current health situation and access to emergency and consular services prior to departure, and be assured that your health and personal security will not be compromised. Make sure that your specific requirements (diet, accessibility, medical care) can be fulfilled before you decide to travel to this destination.	
7 Learn as much as possible about your destination and take time to understand the customs, norms and traditions. Avoid behavior that could offend the local population.	Tourist and local people interactions, Tourist temperament, Language, Contemporary culture.
8 Familiarize yourself with the laws so that you do not commit any act considered criminal by the law of the country visited. Refrain from all trafficking in illicit drugs, arms, antiques, protected species and products or substances that are dangerous or prohibited by national regulations.	Trade in flora, fauna and artefacts, Vandalism, Fire, Drugs and legal matters.

2012 guidebook publications

To better understand the evolution of ethical tourist information in guidebooks the newest editions of these publications in January 2012 (see Box 2011; Miranda et al. 2011; Jenkins 2012) were examined for changes in ethical text. Each book was read in its entirety to compare the overall content with the 2004 versions.

The *Lonely Planet* book on Peru has changed significantly since 2004, with a complete rewrite by new authors and a reorganization of information. Only two ethical statements were found in the introduction section, dealing with tipping and a sense of adventure. Information on culture and history has been placed at the end of the book. The highlighted "Need to Know" section does not include ethical information outside of tipping etiquette. There is less cultural information housed in the regional sections, and mentions of socio-economic information are limited to poverty – though only in the sense of associating this with potential dangers to the tourist (mugging and pick pocketing). Much of the advice to the traveler was related to personal safety rather than responsible behavior.

In the section on Machu Picchu there is no mention of this being a World Heritage Site, and none of the previous environmental issues in the 2004 publication are discussed. Etiquette and ethical behavior on the Inca Trail is also lacking. There is mention of markets that sell whale bones and animal skins, but no discussion of the ethics involved in purchasing these products. The concluding section featured little in the way of ethical text. The information on family travel stressed health issues, but no ethical ones, and the chapter on activities did not discuss any impacts of these on the environment. Again, the crafts and souvenir section did not refer to ethics in the trading of flora, fauna or artifacts though this was addressed under "customs regulations."

The *Rough Guide* is written by the same author, and has retained a similar format to the 2004 version. The introduction section has changed little and still includes ethical information on socio-cultural, environmental and economic issues. Regional sections are also very similar to the 2004 text, with updated information on cultural issues and businesses. In some areas there is a loss of content. For example, ethical issues concerning the purchase of dishes that contain sea turtle meat were mentioned in the introduction to the South Coast area, and again in the restaurant section of the 2004 publication. Now this information appears only in the South Coast introduction section. There is also a lack of specific advice for the tourist. In the concluding section, the 2012 publication discusses climate change and pollution, but does not provide any information about how the tourist could lessen his or her impact on these problems (though there is a short discussion on carbon offsetting in the introduction).

The *South American Handbook* has also retained the same author, and though there is some identical content there are changes in the text, mostly related to updated businesses and pricing. With regard to ethical content, this publication houses a specific section on responsible tourism, though its page extent has decreased since 2004. This two-page section contains updated information on ethical behavior with specific advice listed for the tourist. The "responsible tourism" section is broad in scope; discussing international travel and the benefits of international exchange with local people, touching on issues related to volunteering, ecological footprints, supporting the local economy, bargaining, litter, wildlife, photography and language. The 2012 version also kept ethical information in the regional sections, with some updated wording.

Discussion

Guidebooks present tourists with information that allows them to physically, economically, culturally and environmentally negotiate a destination. In doing so, guidebooks control knowledge through the provision and omission of destination details. In examining the content of the 2004 travel guidebooks, this research demonstrates that these texts are going beyond itinerary, transportation and accommodation suggestions and are presenting information about ethical tourism. This ethical information comprises only a small percentage of material in each book, but it is present.

The presence of ethical information in guidebooks raises other questions related to how guidebook research on ethical information is carried out. Do writers consult the UNWTO or DMOs regarding ethical tourism or is the inclusion of this information based on personal writer preferences and general ethical tourism knowledge? It is unclear how environmental, social and cultural information is chosen. The eight ethical tourist principles outlined by UNWTO are a guide for individuals undertaking national or international travel. In comparing the ethical subject categories from the 2004 guidebooks with these principles, results indicate that guidebooks are providing global ethical tourism information similar to that outlined by the UNWTO. Discrepancies between what information *is* presented and what information *should be* presented to address ethical tourism principles needs to be examined in more detail. The implications of the presentation or omission of destination information in terms of tourist knowledge and appropriate ethical behavior should be further explored to better understand the potential impact on tourist behavior.

In looking specifically at ethical information in the 2004 publications, over a third of ethical statements in guidebooks were worded using persuasive language, indicating that guidebooks are not only aiming to inform tourists, but also to influence tourist behavior. This was done directly through obligatory and advisory statements, and indirectly through optional statements. With these types of statements, the aim is not explanation but an understanding which will elicit an action or activity (Dann 2003). These findings imply that guidebooks seek to shape behavior through messages in the text, reaffirming the idea that guidebooks are actively involved in negotiating and influencing the tourist experience (Lew 1991; Bhattacharyya 1997; McGregor 2000). This leads to questions about the implications of this information for local communities. Guidebooks are advising tourists to either carry out or avoid particular actions. These actions can then affect economic decisions, expectations, social interactions, and tourist understanding of destinations. In attempting to direct and/or influence tourist actions, guidebooks are asserting power over the touristic experience. This potential power needs to be evaluated to understand the implications of these messages in shaping tourist behavior. According to a meta-analysis study on behavioral research carried out by Hines *et al.* (1986/87), the presentation of this type of information can increase the likelihood of more appropriate visitor conduct as knowledge of a problem and actions for change were found to be prerequisites for engaging in responsible behavior. Therefore, presenting persuasive information about ethical behavior that includes explanations of destination-specific problems and tourist-specific actions should increase the likelihood of ethical tourist behavior. However, more research is needed in this area to better understand if and how information translates into changes in behavior. Another issue raised by these findings is that ethical information is scattered throughout the introduction, regional and concluding sections. These are large texts and there is no guarantee that the tourist, having purchased a guidebook, will read all the information. Tourists may skip over sections on the environment or on Peruvian culture and are then not exposed to ethical messages. Therefore, the provision of information in guidebooks, whether persuasive or not, does not necessarily mean that this information will even be seen by the tourist. We do not have enough information on reading patterns associated with these texts. McGregor (2000) sees issues of reception in textual studies as a great challenge in the evolution of cultural research. If this information is read, we do not know if tourists accept or reject guidebook representations of destinations or to what extent tourist interaction with host populations is structured by guidebook information (Bhattacharyya 1997). The incorporation of ideas and messages based on text has been given little attention (McGregor 2000). There has been no general analysis of guidebook cultural history and study on the impact of text on cultures is open for advancement (Lew 1991).

Several researchers have argued that though tourists are a major cause of economic, social and environmental damage, they are shielded from the long-term effects of tourism to the host area and remain ignorant of their impacts (Krippendorf 1987; Moscardo 1996). Based on this argument, tourists should be given more information on ethical tourism and made aware of the impacts of the tourism industry. Pennington-Gray and Thapa (2004) recommend that tourist information providers, such as DMOs, better educate tourists on ethical cultural behaviors. The results of this study indicate that ethical tourism information is being provided by travel guidebooks in a limited way. Not only is information provided, the reader is also being persuaded to adopt particular behaviors and/or attitudes regarding his or her relationship with destination environments, economies and cultures. This pressure from guidebooks is one of the few ways in which individual tourists can be better informed of the potential consequences of their actions at a destination. However, in comparing 2004 and 2012 publications, results indicate that ethical content is waning. The *Rough Guide* showed some ethical content loss in regional sections, but text was very similar to the 2004 version. The *South American Handbook* maintained its "responsible tourism" section, but it was shorter in 2012 than in 2004. This guidebook was the only one that included a specific section addressing ethical choices and behaviors at the destination. The advice was very broad in scope as it related to the whole of South America, rather than Peru specifically, but ethical information was present.

The 2012 *Lonely Planet* update demonstrates a change in content towards a more practical tactical guide for travel. The 2004 edition included similar ethical content to the 2004 *Rough Guide*, with a focus on environmental and sociocultural content. However, the updated 2012 version focuses more on hedonistic pursuits and less on making ethical choices. Even in areas where it would be prudent to mention ethical issues (such as caution about purchasing endangered species after a section discussing markets selling whale bones), this information has been excluded. Though manifest ethical content was limited, an argument could be made that by encouraging visits to cultural and environmental sites the *Lonely Planet* still promotes education, albeit that it relies on the specific sites themselves to provide the educational (and potentially ethical) content.

However, Bhattacharyya (1997) argues that since guidebooks encourage and enhance self-sufficiency they can reduce the tourist's dependence on local populations, so making the assumption that the promotion of local sites in guidebooks will provide for local educational opportunities may be naïve. Access to local perspectives is crucial in providing ethical representations of a destination (Tegelberg 2010). Therefore, by including ethical tourism information, guidebooks should be clearly and openly encouraging interaction with local peoples. This type of advice is present in the 2012 *South American Handbook* and *Rough Guide*, but has been curbed in the updated *Lonely Planet*.

In examining guidebook content, it is important to keep in mind that the main goal of guidebook publishers is to sell guidebooks. Guidebooks are designed to explain the attractions at a destination and provide information allowing for independent travel. However, these texts are participants in the tourism industry and are therefore subject to examination in terms of their role and potential impact on tourists and local communities. The goal of this research is to evaluate the ethical content of guidebooks, to better understand the information being given to tourists, and the potential role that guidebooks play in presenting this information. This study demonstrates that guidebooks do contain a limited amount of ethical content on destinations but that this content (and its focus on ethical behavior) may be in decline.

Conclusion

Findings and discussions based on this research indicate that the role of the guidebook is multi-dimensional. Guidebooks provide and assess travel information (activities, accommodation, transportation, etc.) that helps tourists to negotiate a destination. With this assessment of information, guidebooks also act as an educative tool, evaluating tourist sites and travel routes. Through the inclusion and omission of information, guidebooks have the potential to exert a massive influence on the tourist experience (McGregor 2000). This position, juxtaposed between host and guest, allows guidebooks to negotiate the relationship between tourists and destinations. This is a powerful position, and one that has the capacity to influence the travel practices of readers. The *Lonely Planet* is an influencing series (Tegelberg 2010), and these types of series have the potential to act as a source of ethical information and a promoter of ethical tourist behavior. Results of this study indicate that these texts are addressing destination-specific ethical issues and promoting ethical travel principles in a limited way. However, this study also found that ethical content changed between the 2004 and 2012 versions of each series. A more thorough analysis of changes in ethical information in guidebooks over time is warranted. In 2004 guidebooks were found to be engaging in ethical tourism through text, and therefore potentially educating the reader on his or her impacts and responsibilities at a destination. In 2012 ethical content did not increase in any of the publications.

So, eight years on, the presentation of ethical information to potential tourists is not advancing in guidebooks and is in fact declining in some of the publications. Is the decrease of ethical content found in the *Lonely Planet* a trend for this series? If so, this trend is worrying. Are guidebooks, keen to maintain relevance in a changing market, pursuing a different style of information presentation which omits advice and rules on ethical behaviors and responsible choices at destinations in favor of a more streamlined tourist-centered hedonistic approach? Is ethical tourism itself becoming an endangered issue? Iaquinto's (2011) research on guidebooks from the perspective of the writers reveals a frustration with the editorial controls which appear to be directed towards a more mainstream tourist. This may have a bearing on ethical tourism content if the series seeks to concentrate on providing more avenues for mainstream tourist entertainment and less on providing an alternative tourism education. More research on publication choices in ethical information is needed.

Based on the results of this study much more research is needed to gain a better understanding of ethical information being presented to tourists. First, are guidebooks presenting destination-appropriate ethical information? The 2004 guidebook information examined in this study does conform to the UNWTO's global Responsible Tourist Principles, however these are very general codes of ethics which may or may not reflect the current needs of specific destinations. Future studies should involve research on destination-specific ethical behaviors and compare these with the information given in popular sources to determine whether or not the information is compatible with the concerns of the particular destination in question. Second, how is the ethical information studied and interpreted by the guidebook reader? Future studies in this area should examine how guidebooks are read and how ethical information included in them is processed, interpreted and retained by the reader so that findings can be incorporated into ethical information design and dissemination. Third, does ethical information lead to changes in tourist behavior? Though this research has shown that some ethical statements in

Ethical information in travel guidebooks 185

guidebooks have persuasive language, it is still unclear as to whether or not persuasive messages have the ability to influence behavior. More empirical research in this area would help to determine the kinds of information-related variables that can help or hinder behavioral change in tourists, giving the industry better tools to encourage ethical tourism.

Guidebooks, as part of the tourism industry, should be working towards sustainability. Concerns regarding the need for ethical tourism have been expressed on an international level by UNWTO. In recognizing the role of the guidebook as a marketing and educative tool, these texts need to be evaluated in terms of how they address ethical tourist issues. Guidebooks offer an avenue for presenting information on tourist impacts directly to the tourist. This study concentrates on the current nature of the relationship between guidebooks and ethical tourism. It is also one of a small number of academic analyses which focuses on how information sources communicate ethical tourism principles. More research in this area is needed to fully understand the current and potential role of guidebooks in ethical tourism.

Activity

Examining ethical information in travel guidebooks

Choose a destination you have visited or would like to visit in the future. Do some research on the destination, including looking at the destination's official tourist site, to figure out the popular tourist activities taking place there and the tourism infrastructure that provide for those activities.

Once you have a better idea of what is happening at the destination, brainstorm some of the negative impacts of tourism that are or could be a problem. Write these impacts in point format in column A of the table below.

Now brainstorm some of the potential ethical tourist behaviors that can address these impacts. Use the sources in the further reading section below to help with this. Write the possible positive ethical tourist behaviors related to each tourism impact in column B of the table below.

Table 11A

A. Negative impacts of tourism	B. Potential ethical tourist behaviors to address this
Example: High number of plastic water bottles left on beach	*Example:* Encourage use of reusable drinking bottles

If you were a destination marketer in charge of encouraging ethical tourism in your area, what sort of information would you concentrate on? How would you encourage ethical tourist behavior? Create a draft of a brochure on ethical tourism that would address specific ethical concerns and behaviors for your chosen destination.

Further sources

Bhattacharyya, D.P. (1997) "Mediating India: An Analysis Of A Guidebook" in *Annals of Tourism Research*, 24 (2): 371–89.

Global Code of Ethics for Tourism. Online, available at: http:www.unwto.org/ethics/responsible/en/pdf/brochure_e.pdf (accessed January 7, 2013).

McGregor, A. (2000) "Dynamic Texts And Tourist Gaze: Death, Bones And Buffalo" in *Annals of Tourism Research*, 27 (1): 27–50.

Responsible Travel: www.responsibletravel.com/copy/responsible-tourism.

Tourism Concern: www.tourismconcern.org.uk/ethical-tourists.html.

References

Babbie, E. (2001) *The Practice of Social Research*, Belmont, CA: Wadsworth/Thomson Learning.

Bhattacharyya, D.P. (1997) "Mediating India: An Analysis Of A Guidebook" in *Annals of Tourism Research*, 24 (2): 371–89.

Box, B. (2004) *South American Handbook 2005* (81st edition), Bath, UK: Footprint HandBooks.

Box, B. (2011) *South American Handbook 2012* (88th edition), Bath, UK: Footprint Handbooks.

Carter, S. (1998) "Tourists' And Travellers' Social Construction Of Africa And Asia As Risky Locations" in *Tourism Management*, 19 (4): 349–58.

Dann, G. (2003) "Noticing Notices: Tourism To Order" in *Annals of Tourism Research*, 30 (2): 465–84.

Gilbert, D. (1999) "'London In All Its Glory – Or How To Enjoy London': Guidebook Representations Of Imperial London" in *Journal of Historical Geography*, 25 (3): 279–97.

Goodwin, H. and Francis, J. (2003) "Ethical And Responsible Tourism: Consumer Trends In The UK" in *Journal of Vacation* Marketing, 9 (3): 271–84.

Hines, J., Hungerford, H. and Tomera, A. (1986/87) "Analysis And Synthesis Of Research On Responsible Environmental Behavior: A Meta Analysis" in *The Journal of Environmental Education*, 18 (2): 1–8.

Iaquinto, B.L. (2011) "Fear Of A Lonely Planet: Author Anxieties And The Mainstreaming Of A Guidebook" in *Current Issues in Tourism*, 14 (8): 705–23.

International Centre for Responsible Tourism Canada (ICRT Canada) (2001) *What Is Responsible Tourism?* Online, available at: www.icrtcanada.ca/?page_id=27 (accessed December 14, 2011).

Jenkins, D. (2003) *The Rough Guide to Perú*, New York: Rough Guides.

Jenkins, D. (2012) *The Rough Guide to Peru*, London: Rough Guides Ltd.

Kosher, R. (1998) "'What Ought To Be Seen': Tourists' Guidebooks And National Identities In Modern Germany And Europe" in *Journal of Contemporary History*, 33 (3): 323–40.

Krippendorf, J. (1987) *The Holiday Makers: Understanding the Impact of Leisure and Travel*, Oxford, UK: Heinemann Professional Publishing.

Lew, A. (1991) "Place Representation In Tourist Guidebooks: An Example From Singapore" in *Singapore Journal of Tropical Geography*, 12 (2): 124–37.

McGregor, A. (2000) "Dynamic Texts And Tourist Gaze: Death, Bones And Buffalo" in *Annals of Tourism Research*, 27 (1): 27–50.

Mehmetoglu, M. and Dann, G. (2003) "Atlis/Ti And Content/Semiotic Analysis In Tourism Research" in *Tourism* Analysis, 8 (1): 1–13.

Miranda, C., Dowl, A., Shorthouse, K., Waterson, L. and Williams, B. (2011) *Lonely Planet: Discover Peru*, Lonely Planet Publications Ltd.

Moscardo, G. (1996) "Mindful Visitors: Heritage And Tourism" in *Annals of Tourism Research*, 23 (2): 376–97.

Osti, L., Turner, L.W. and King, B. (2009) "Cultural Differences In Travel Guidebooks Information Search" in *Journal of Vacation Marketing*, 15 (1): 63–78.

Otness, H. (1993) "Travel Guidebooks: A World Of Information For Libraries" in *Wilson Library Bulletin*, 67 (5): 38–40 and 116–17.

Pennington-Gray, L. and Thapa, B. (2004) "DMOs And Culturally Responsible Behaviours: An Exploratory Analysis" in *Tourism: An Interdisciplicary International Journal*, 52 (2): 183–94.

PromPerú (2003) *Perfil del Turista Extranjero 2002*. Online, available at: www.Perú.org.pe/Publicaciones.asp (accessed April 30, 2004).

Rachowiecki, R. and Beech, C. (2004) *Lonely Planet: Peru*, Footscray, VIC: Lonely Planet Publications.

Santos, C.A. (2004) "Framing Portugal: Representational Dynamics" in *Annals of Tourism Research*, 31 (1): 122–38.

Tegelberg, M. (2010) "Hidden Sights: Tourism, Representation And Lonely Planet Cambodia" in *International Journal of Cultural Studies*, 13 (5): 491–509.

United Nations World Tourism Organization (UNWTO) (2001) *Global Code of Ethics for Tourism*. Online, available at: www.ethics.unwto.org/en/content/global-code-ethics-tourism.

United Nations World Tourism Organization (UNWTO) (n.d.) "The Responsible Tourist and Traveller" in *Global Code of Ethics for Tourism*. Online, available at: www.unwto.org/ethics/responsible/en/pdf/brochure_e.pdf.

Wong, C.K.S. and Liu, F.C.G. (2011) "A Study Of Pre-Trip Use Of Travel Guidebooks By Leisure Travellers" in *Tourism Management*, 32 (3): 616–28.

12 Business travel and the environment

The strains of travelling for work and the impact on travellers' pro-environmental *in situ* behaviour

Wouter Geerts

Introduction

The environmental sustainability of hotel operations is increasingly scrutinized by hotel management and other stakeholders. Because of this intensified scrutiny, as well as higher costs associated with unsustainable activities, hotels are progressively implementing and adopting environmentally friendly operations (Miao and Wei 2013). Making adjustments to the hotel's construction, introducing new technologies, and implementing more efficient energy, water and waste management systems are common practices introduced by a growing number of hoteliers (Hawkins 2006; Budeanu 2007; Tierney 2007). Kirk (1995) found that hotels are mainly implementing such practices for financial gain, while Tzschentke *et al.* (2004) further found that small accommodation owners also implemented sustainability practices as part of a moral obligation to reduce their environmental impact. Although the above research has shown that implementing environmental practices can be beneficial to hotels' bottom line, findings have been less conclusive on the effect of these types of practices for booking intentions and guest satisfaction.

Research that aims to provide an insight into hotel guests' attitudes towards environmental practices predominantly focuses on one of two issues: whether potential guests book hotels because of sustainability practices, and whether guests who stay take part in sustainability practices. First, findings about a willingness to pay extra for 'green' hotel stays are inconclusive, with some research supporting guests' willingness to pay extra (e.g. Clausing 2008), other research finding an unwillingness to pay extra (e.g. Millar and Baloglu 2011), and yet other research linking willingness to guests' predetermined environmental concerns or financial means (Kang *et al.* 2012). Second, research into environmental behaviour by hotel guests is limited, with a main contribution by Goldstein *et al.* (2008) in relation to towel and linen reuse, and more recently a noteworthy effort by Miao and Wei (2013), researching the differences between environmental behaviour in both home and hotel settings. Shortcomings of these works, however, can be found in the narrow and segregated focus on the implementation and uptake of environmental practices without significant consideration for external influences.

This chapter will discuss how these 'external influences' impact the pro-environmental behaviour of business travellers. The reason for researching business travellers is twofold. First, since the large majority of research at the intersection of tourism and sustainability is aimed at leisure travellers, business travel has often been neglected; this chapter aims to contribute to the growing body of literature and to improve our understanding of business travellers' environmental behaviour. Second, due to differences in the purpose of undertaking trips, business travellers' attitudes towards travelling, and their usage of hotel facilities, are expected to be different from those of leisure travellers. Business travellers are particularly suited for the discussion in this chapter, because work duties – an external influence – are expected to impact travellers' behaviour. The aim of this chapter, then, is to discuss how the *in situ* (i.e. at the destination) lifestyle of business travellers – which often involves short trips to different locations and time zones to perform stressful work duties – impacts travellers' environmental behaviour. After a brief discussion of literature on business travel, and an explanation of the methodology, the discussion will first focus on travellers' attitudes to travelling and staying in hotels, followed by a discussion on the impact of these attitudes on environmental behaviour. The chapter will conclude with some remarks and an assessment of implications for future research.

Conceptualizing business travel

In this chapter, the usage of the term 'business traveller' is based on the definition provided by Davidson and Cope (2003: 3): '[I]ndividual business travel comprises the trips made by those whose employment requires them to travel in order to carry out their work.' To define business travel more comprehensively, a comparison with leisure travel is helpful. While boundaries between business and leisure travel can become blurred (e.g. when business travellers take part in leisure activities while on a business trip), there are some differences between leisure and business trips which have an effect on travellers' behaviour.

The first difference is the destination: while leisure travel has a wide variety of places in which it takes place, like nature-based holidays, city breaks, or coastal resorts, business travel most often will only take place in urban areas. Hence, business trips are generally characterized as 'movement[s] between locations, consisting of airports, office buildings, and hotels, (...) giving the trips a monotonous character' (Lassen 2006: 307). Lassen (2006, 2009, 2010) refers to business travel as a 'life in corridors', with cultural differences between destinations being erased because of the 'cocooned passage' through these locations (McNeill 2008). According to McNeill (2008), business hotels are monotonously styled to conform to the expectations of guests. Business travellers want predictability, a certain standard of service, and an anonymous form of hospitality, which incidentally leads to many hotels looking and functioning in a similar fashion (Bell 2007).

A second difference between leisure and business travel relates to the decision-making process behind a trip. For business travellers the employer often decides and pays for travel (Swarbrooke and Horner 2001; Davidson and Cope

2003), while leisure travellers will generally book and pay themselves. The notion of employers making decisions on travel for their employees has a number of consequences, of which two are especially important to this chapter. First, because employees travel to satisfy their employers' expectations, such travels can be experienced as strenuous and negative, especially when having to undertake a high number of short trips, an activity referred to as 'hypermobility' (Becken 2007). Travelling in business class or collecting loyalty points (for private use at a later date) can increase personal satisfaction, but might contravene company policies (Douglas and Lubbe 2009, 2010). Importantly, the strain of business travel not only affects the traveller, but can also impact his or her family (Ivancevic *et al.* 2003; Gustafson 2006). Business travellers will generally travel alone, having to separate from family and friends. Gustafson (2006) found that present-day business travel is a source of conflict for business travellers between obligations to work and obligations to the family.

The second consequence of employees' travelling behaviour being (partially) ruled by employers can be situated at the intersection of business travel and environmental behaviour. Lassen (2010), in his research into business travellers' environmental considerations, found that individual business travellers generally would not consider the environment when travelling. Time, money and comfort were seen as more important, with air travel being perceived by many as a given that could not be affected or changed. With employers making decisions on travel instead of individual travellers, employees might feel powerless to reduce their environmental impact, or might use powerlessness as an excuse for inaction. This chapter will provide an insight into the impact of these factors on environmental behaviour.

Methodology

Research to understand business travellers' behaviour was undertaken in London, United Kingdom; a global hub which attracted just over 20 million international visitors in 2011 (*The Independent* 2011) and hosts an estimated 101,000 hotel rooms (Greater London Authority 2002a). Business travel contributes £2.75 billion to the city's economy (Greater London Authority 2002b) and it is argued that more than 50 per cent of hotel rooms are occupied by business travellers (Davidson 1994).

The research undertaken for this chapter was part of a broader research project into business travellers and pro-environmental behaviour. In order to research travellers' attitudes towards environmentalism, and to assess their actual behaviour when travelling, the focus of the research was on their *in situ* behaviour (i.e. their behaviour at the destination). Because there has only been limited research into business travellers and their environmental considerations, especially relating to the *in situ* aspect of travelling, the exploratory nature of research resulted in the application of grounded theory analysis (Glaser and Strauss 1967) throughout the research process. A total of 34 semi-structured interviews with business travellers and six three-hour long observation sessions were carried out, which provided the majority of the data discussed in this chapter. To enhance the reliability of the interview data, a further 22 London hotel managers were interviewed, managing hotels ranging from budget class to five-star. Furthermore, four interviews with company representatives were carried out so that information could be triangulated and claims from business travellers could be checked and compared.

Three methods to recruit business travellers for interview were successfully used, starting with purposive sampling and snowballing, which generated 18 interviews lasting between 45 and 60 minutes. Second, a hotel manager of a five-star hotel in central London was contacted to request access to his guests. He offered to email some of his regular business guests, asking for their participation. The email introduced the research and affiliations. This resulted in four 45–60 minute interviews, which all took place in the bar of the hotel. Third, in an attempt to widen the scope of interviewees, London City Airport was contacted to ask permission to approach travellers.

London City Airport was selected because it is a relatively small airport that is closely located to both Canary Wharf and the City, the two major business districts in London. It operates regular flights to cities across Europe and business-class-only flights to New York, and prides itself with having the shortest check-in times of any London airport. The airport is heavily focused on the business traveller, with 64 per cent of all travellers using London City Airport doing so for business purposes (London City Airport 2011). Permission was granted for two days, but only for the landside[1] of the airport. During the two days, 11 interviews were undertaken. The interviews were relatively short – about 10 minutes on average – as people either had to catch a flight or were rushing into Central London to carry out their business. Only a restricted number of questions could be asked, with the questions largely focusing on the interviewee's choice of hotel and their environmental behaviour. Because of saturation of the collected data, it was decided to stop recruiting business travellers once 34 were interviewed. This is in accordance with Glaser and Strauss's (1967) discussion on saturation, in which they point out that the researcher should judge when new empirical data is not adding any new insights into the issues under investigation. For the purpose of anonymity, all interviewed business travellers are solely referred to by an ID number, as shown in Table 12.1. All interviews were recorded, transcribed and uploaded to qualitative analysis software (Atlas.ti 6.1) and analysed using Schmidt's (2004) coding guide, which resulted in a number of key themes. Two of these, the negative attitude to travelling and the low levels of environmental considerations, will be the focus of the discussion in this chapter.

Next to interviews, business travellers were observed in the hotel setting on six separate occasions. Interactions between business travellers and hotel staff at the reception desk were considered as potentially interesting, so in two sessions observations took place in the hotel lobby. A further four observation sessions took place in hotel bars. All sessions took place in different hotels, with two hotels situated in the Canary Wharf business area, and four in or close to the City of London. In some cases travellers were approached for informal conversations. During the participant observation sessions, data was recorded in the form of field notes. As

Sunstein and Chiseri-Strater (2002: 56) note, 'the difference between doing fieldwork and just "hanging out" is the writing'. For that reason, a laptop was taken to the sessions and full field notes were taken while the observation was taking place.

Findings and discussion

Travellers' attitudes towards travelling

Attitudes towards travelling for work differed considerably among travellers. A range of aspects influenced these attitudes, with the nature of, or reason for,

Table 12.1 Business travel interviewees

ID	Gender	Age bracket	Nationality	Industry	Business trips per year
1	Male	35–44	Canadian	Oil & Gas	4
2	Female	25–34	Canadian	Oil & Gas	1
3	Male	25–34	German	Telecommunication	30
4	Female	25–34	British	Consultancy	3
5	Male	45–54	British	Tourism & travel	12
6	Female	45–54	British	Law	16
7	Male	25–34	Canadian	Consultancy	10
8	Male	35–44	Spanish	Telecommunication	100+
9	Male	35–44	Greek	Academia	12
10	Male	55–64	British	IT	55
11	Male	45–54	British	Consultancy	60
12	Male	35–44	British	Manufacturing	15
13	Male	25–34	Canadian	Finance	3
14	Male	45–54	Dutch	Tourism & travel	12
15	Male	25–34	German	Consultancy	24
16	Male	55–64	Dutch	n/s	40
17	Male	45–54	Irish	IT	10
18	Female	35–44	French	Consultancy	30
19	Male	65+	British	Tourism & travel	20
20	Male	35–44	US American	IT	75
21	Male	35–44	US American	Telecommunication	20
22	Male	55–64	British	Law	50
23	Female	45–54	Belgian	Oil & Gas	12
24	Male	35–44	British	Manufacturing	45
25	Male	55–64	Austrian	Manufacturing	20
26	Male	25–34	Canadian	Telecommunication	20
27	Male	35–44	German	Consultancy	50
28	Male	35–44	British	Construction	10
29	Female	35–44	Canadian	Third sector	12
30	Male	35–44	British	Media	3
31	Male	25–34	British	Third sector	3
32	Female	35–44	Vietnamese	Finance	5
33	Male	45–54	Indian	Finance	8
34	Male	35–44	French	Finance	6

travelling, and the frequency of travel, heavily impacting travellers' attitudes. Although business travellers are often referred to as a homogenous group, there are many reasons for individuals to travel for work, and it seems that these differing reasons impacted the travel experience of the interviewed travellers. The research sample consisted of consultants who for periods of up to a few years travelled every week to the same location, only being home during the weekend. A German consultant, for example, had for the past year travelled every Sunday evening to London, returning home on Thursday evening. These consultants effectively had to travel long distances to get to their offices, and hence stayed in a hotel during the working week.

This is in stark contrast with other travellers, especially those who travelled to meet clients or who undertook intrafirm travel (Faulconbridge *et al.* 2009). Their trips were generally shorter and less frequent, but often also conflicting with their day-to-day jobs back home. This meant that consultants often had spare time in the evening, while other travellers would often work both during the day time and at night – tasks relating to the reason for travelling during the day, and other day-to-day tasks (like checking emails, video-meetings with superiors etc.) at night. Finally, there were travellers who undertook trips to varying destinations because travel was part of their day-to-day job. Among these travellers was a sales representative for a paper manufacturer, who said he travelled 40 weeks a year selling his employer's goods; and a motivational speaker and course leader who visited companies around the world to convene internal training sessions. These travellers generally had considerable spare time in the evening, like the consultants, but did not visit the same location on a regular basis. Due to these differences, most consultants said that travelling was 'boring and mundane', while other travellers often mentioned it was stressful and tiring.

Frequency of travel was another aspect that was found to impact upon travellers' attitudes towards travelling. Some interviewees travelled irregularly (say once every two months or less) and were generally positive about their travelling experiences, or as one traveller (ID 2) put it more aptly, she was 'not sick of it yet'. In contrast, other interviewed travellers stated that they were away from home for around 200 nights per year. Some were consultants spending more than 40 weeks in the same location, while one traveller had over 100 short (up to two nights) trips per year to a range of locations. It was these – arguably more experienced and hardened – travellers who often had a negative opinion of their travel commitments, as the following quote from a Canadian account manager illustrates:

> You can have different meetings in different locations, so you are city hopping, and that is very extreme. Then you can fly in somewhere in the morning, have a meeting in the afternoon, and fly to the next location during the evening. I don't really like it; it's very tiring and extreme. Normally I set myself a goal of seeing at least one attraction, landmark or museum when at a new location. That is often impossible when city hopping.
>
> (ID 7)

This 'hypermobile' travel behaviour is, according to Becken (2007), a growing phenomenon among leisure travellers (more and shorter trips throughout the year), and something that is strived for by many, while interviewed business travellers were more negatively disposed to the extremity of the experience, possibly because for business travellers hypermobility constitutes an even higher frequency and shorter stays than the leisure travel equivalent. Furthermore, as the above traveller mentioned, while leisure trips can predominantly consist of visiting 'attraction[s], landmark[s] or museum[s]', business travellers will arguably spend most of their time in aeroplanes, hotels and 'corridors' (Lassen 2006). The following quote from a director of a telecoms provider illustrates his 'life in corridors', and how it influences his travel experience:

> When I first started I was really excited, because I [thought] 'ooh I get to see the world, travel all over the place', but you learn pretty quickly that the insides of cabs, hotels, airplanes, airports and boardrooms all look the same no matter where you are in the world. So it starts off as this exciting curiosity, but now to me it's like a massive inconvenience, and usually [I'm] tired and [have] jetlag.
>
> (ID 26)

As the interviewee notes, business travellers generally spend most of their time in aeroplanes, airports, offices and/or hotels. These spaces are characterized by the large amount of people moving through them, without 'living' in the space, and are otherwise referred to as non-places (Sheller and Urry 2006; Urry 2007, 2009; Lassen 2009), and liminal, or in-between, places (Pritchard and Morgan 2006). Lassen (2009) illustrates the idea of non-places by referring to travellers waking up in a hotel room and having to take some time to think where in the world they are, because all hotel rooms look similar. Cultural differences in non-places are erased because of the 'cocooned passage' through these spaces (McNeill 2008). Hence, international work trips are characterized by 'movement between locations, consisting of airports, office buildings, and hotels, (…) giving the trips a monotonous character' (Lassen 2006: 307), which negatively influenced the interviewed travellers' attitude towards travelling.

As the above quotes have already made mention of, one of the main negative attributes of regular business travel is the stressfulness. Twelve travellers of the total sample of 34 interviewees mentioned that travelling is a tiring experience, as the following quote from a financial research associate illustrates:

> The reason that I stopped liking [my previous job] was the amount of travel.…You can't get into any life routine, and it starts to take a toll on health as well, because you can't keep up on sports hobbies, so I couldn't go mountain biking for all that time because I was travelling so much, and you're eating out every day, so you go to restaurants and [are] just getting fat basically.
>
> (ID 13)

The interviewee says that by being away it is harder to keep to a healthy diet and to find the time to exercise, which results in a strain on his physical health. Four other travellers similarly made the explicit link between travelling and their health. As Ivancevic *et al.* (2003) argue, travelling can be a stressful activity due to travel delays, heavy workloads, the feeling of loneliness and concerns about personal security. Living highly mobile lives can be detrimental to an individual's wellbeing, with the possibility of suffering depression, anxiety and emotional disconnection (Elliott and Urry 2010). Stresses are apparent before and after travel, and the impact of travelling is not solely upon the traveller. Preparing for a trip, and dealing with feelings of guilt after return, can be equally stressful, especially when leaving a family (Ivancevic *et al.* 2003). With the exception of one traveller whose children were grown up, all 17 travellers with a family noted that they were conscious about the impact their travelling had on their family life, and that they were trying to reduce this impact. The following is a quote from a Greek traveller who made travelling decisions based on his home life:

> Because I have a young family it isn't easy for me to travel to conferences in Australia or North America. Not impossible, but my wife is heavily pregnant, [and] I don't want to be on the other side of the world when something happens.
>
> (ID 9)

It can be argued that family would have been less of a consideration in the past. Traditional societal expectations of work division between men and women, with men in business and women in the household, are changing (Harris and Ateljevic 2003). In the eighteenth and nineteenth centuries, travellers were mostly men, portrayed as 'explorers' and 'adventurers' (Harris and Wilson 2007). With a growing number of women travelling for business (Harris and Ateljevic 2003; Harris and Wilson 2007), however, Gustafson (2006) found that present-day business travel for many families is a source of conflict between obligations to work and obligations to the family. While men travel considerably more than women, regardless of their family situation (Gustafson 2006), as the above quote illustrates, most interviewed male travellers were concerned about leaving their family frequently.

Although positive attributes of travelling were scarce in comparison to negative attributes, some interviewees mention aspects that made travelling more bearable or enjoyable. Five travellers made mention of their attempts to have a holiday after a business trip. This is a cheaper alternative to buying separate flights and, in contrast to a 'cocooned passage' through spaces (McNeill 2008), being a leisure traveller brings the benefit of free time and the freedom to explore. This is often impossible as a result of the high workload when travelling for business reasons. With such a high workload, however, the following German traveller thought that being on an aeroplane could be beneficial:

> Two hours on a plane is a big luxury, because it is two hours where you can actually have undisturbed time to do your email....No internet, no phone, no new emails, so literally you can focus for two hours, so that's not entirely a bad thing.
>
> (ID 3)

This relates to Solomon's (2009) argument that many individuals enjoy their daily commute between home and work. She bases her discussion on work by De Grazia (1962), who argued that to his New York respondents their daily work commute was a relaxing activity. Although written 50 years ago, findings in this chapter show De Grazia's findings could still be relevant; the above quote illustrates how spending two hours in a secluded space, like an aeroplane, can be used to catch up on work – using the aeroplane as an office – and beneficial to travellers' attitude to travelling. While Solomon's (2009) discussion focuses exclusively on commuting to work, the discussion can be extended to business travel, and as the following quote from a 'hypermobile' Spanish traveller shows, not just to the travelling aspect of business travel,

> Given that we just had a second child....you would want to spend an amount of time at home. Having said that, the situation at home is also really, really busy, so to be honest, sometimes the overnight stay at the hotel could be a way to catch up on work, catch up on sleep.
>
> (ID 8)

To this interviewee, with a young family, travelling is beneficial as his time apart from the family means that he has fewer distractions. As Solomon (2009: 166) discussed in relation to commuting, '[I]t is the time for unwinding before facing the family.' While being away from family is a reason for some travellers to consider changing jobs, the above traveller turns his periods of absence into a positive aspect of travelling. These differing attitudes towards travelling show that business travellers are a heterogeneous group of individuals, with different experiences and differing attitudes towards these experiences. The frequency and length of travel, the family situation, the locations visited, and the workload are important factors that influence travellers' attitudes. With some caution, however, it could be argued that most travellers are more inclined to point out the negative aspects of travel over the positive ones. In research referred to in the above sections, the analysis often stops with this notion; however, this chapter will extend beyond this, and aims to investigate how travellers' negative attitudes to travelling influenced their behaviour at the travel destination. With travellers predominantly staying in hotels while away from home, attitudes towards these hotel stays will first be discussed.

The importance of the 'fixed' hotel in a transient lifestyle

Hotels are by far the most-used form of accommodation by business travellers. Of all visiting international business travellers to the UK in 2009, 75 per cent stayed in a hotel or guest house (ONS 2011). Indeed, all interviewed travellers had considerable experience of staying in hotels, but most travellers had less of a strong view of hotel stays than of the total travel experience. Answers like 'most places are adequate' (ID 8), 'at the end of the day it's a bed' (ID 13) and 'it's normally fine' (ID 26) provide an indication of the general attitude of the interviewees towards hotel stays. Instead of discussing their attitudes towards staying in a hotel, most travellers would discuss the importance of certain amenities or services, often along the lines of 'as long as it has got an internet socket' (ID 4, also mentioned by ID 7).

Other important amenities and services were 'a comfortable bed' (ID 12, 13, 14, 25), 'proper breakfast' (ID 14, 20), 'priority check-in' (ID 3) or 'a clean room' (ID 4, 25). These rather apathetic answers could be understood as conferring low importance on hotel quality or the hotel experience. But it could instead be argued that such responses indicate how important a hotel is to travellers, and how travellers expect hotel stays to be hassle-free and providing a service that simplifies their lives. Therefore, any hotel practices that conflict with these expectations will be referred to as 'friction' to a traveller's hotel stay.

Major hotel chains generally have building and service standards that render all their hotels similar in terms of exterior and interior design, amenity and service, and ambiance and staff etiquette. Furthermore, with star-rating schemes, standards are guaranteed to be similar for hotels with the same rating. As a result, hotel rooms from Hilton, Marriott or Intercontinental are very similar, both within the brand and across brands, except for some design features. Hence, it is 'special touches', the seemingly insignificant aspects of service like, for example, a hotel chain that 'remember[s] what kind of pillows you like' (ID 13), that make the difference for many travellers.

As most of the interviewed travellers stayed in four- or five-star hotels, a comfortable bed or luxurious bathroom were *expected* and generally a given, while an extra pillow, priority check-in, or a panoramic view from the window were *desired*. An extra pillow might seem an insignificant service for infrequent travellers, but for regular travellers it is these 'special touches' that can make a loyal customer, because it means that the hotel stay constitutes less 'hassle' and friction. Indeed, most of the regular travellers were very loyal customers to 'their' hotel brand. The following quote from a British consultant gives some insight into the reasons behind loyalty,

> I always pretty much stick with Marriott, I like their formula.... You know what you're getting, and when you're working you don't have to worry about that sort of thing. One of my colleagues said to me once when we were in New York [that] he was sick of the [chains], so he was going to a boutique hotel. He booked it and it was shit; I said 'it serves you right', you know, at least if you book somewhere that [is a chain], you have consistency.
>
> (ID 11)

This traveller tries to stay in Marriott hotels whenever he travels for work, because he knows what to expect. This notion of loyalty and the need for certainty and predictability is interesting when considering the negative attitude many travellers had towards the mundane and homogeneous character of business travel. Surely one would not expect the same travellers that complain about the mundane nature of travelling to be trying to increase the predictability of the trip by always choosing the same hotel. In fact, this is exactly what the interview data shows. Instead of switching brands, or going to a boutique hotel, regular travellers prefer to know what they can expect.

When travelling to another country or continent which they have not visited before, these travellers still decide to stay in one of the large 'Western' hotel chains because they know what to expect. The existence of websites like TripAdvisor is predominantly based on the uncertainties resulting from travelling to unknown places, and a common distrust of hotels' marketing and PR materials. The pursuit of a predictable hotel stay is part of what Ritzer (2004) refers to as the McDonaldization of society, with predictability and standardization important aspects of the contemporary Western consumption culture. Brand loyalty provides upgrades and extras, but it also reduces stress and ensures travellers can focus on the job, both before the trip and after arrival.

Hotels, then, play an important role in 'fixing' mobile bodies (McNeill 2008). To use Lassen's (2006) terminology, hotels are part of the 'monotonous corridors' travellers spend most of their time in. The monotonous character of hotel spaces can lead travellers to argue that their hotel stays are 'boring and mundane', but it also results in a predictability which is often desired by those same travellers. This desire is particularly evident when considering the brand loyalty of regular travellers. Some interviewed travellers, predominantly consultants regularly visiting the same hotel, went on to highly rate the importance of social relationships and interactions with hotel staff. Although commercial hospitality is argued by some to be 'fake' (Brotherton and Wood 2000; Lashley 2000; Lockwood and Jones 2000; Ritzer 2007), and hence relationships made with staff or owners of these commercial hospitality establishments equally fake, for some regular and hypermobile travellers these 'fake' relationships with hotel staff were important. While the knowledge that hotel staff are paid to be friendly and courteous towards guests left many travellers unwilling to partake in 'fake' relationships with them, for some these relationships with hotel staff were more important. Remember how the traveller in the quote above stated that he liked Marriott's 'formula'. When asked what he perceived this to be, he referred to the friendliness of the staff. Another traveller, a consultant always visiting the same London hotel, said,

> When you walk in you're welcomed. Everybody says 'hello' to you 'hello Mr. (name of interviewee)'. Even if they don't know you [they say] 'good evening sir'. You're made welcome....In the [business] lounge there's an old lady called Betty, she looks after the lounge. We get on like a house on

fire, we have a good relationship. I welcome that comfort and warmth that I get from the people here.

(ID 10)

Living the mobile life of a regular traveller often means inhabiting or visiting large numbers of spaces in a short time. Many of these spaces are not fixed (like aeroplanes), are transient, and involve little contact with other individuals. This isolation from social contacts is intensified by the large workload of many travellers, and face-to-face interactions are reduced by the proliferance of mobile phones and laptops.[2] The hotel, then, is often one of the few places that is 'fixed', and where human face-to-face contact is 'offered'. Human contact can be found in the hotel bar, the restaurant or the more exclusive business lounge, where travellers can interact with hotel staff. Contrastingly, if and when travellers are looking for solitude they have the possibility of retreating to their room, and, as increasingly introduced by hotels, use check-in and check-out services which require no human interaction. As already pointed out, some travellers enjoyed interacting with staff and fellow guests during their hotel stay, while many others expected their hotel stay to be one with minimal interaction. These latter travellers desired hotels to maintain, as McNeill (2008: 391) argues, an 'anonymous, commercially understood form of hospitality'. As will be discussed in the following section, these preferences concerning social interaction (as well as the strain of travelling) often have a negative impact on travellers' environmental considerations and behaviour.

Business travellers' environmental in situ *behaviour*

When asked about environmental sustainability and climate change, the large majority of travellers (30 of 34 interviewed travellers) accepted that climate change is happening and that human actions contribute to this process. The remaining four interviewees did not deny the existence of climate change, but argued that it was a natural phenomenon that could not be impacted by human behaviour; as one interviewee (ID 25) argued: '[M]en will always be able to destroy men, but not the Earth.' The large majority of travellers, then, agreed that climate change is an unwanted phenomenon with possible disastrous outcomes and that mitigation is needed. Furthermore, many noted that their travelling lifestyles were not always in line with strategies to mitigate human impacts on the environment. One regular long-haul traveller, however, argued the following,

> I certainly believe in the impact [that] particularly our generation and previous generations have had on the planet. But it's funny, I never really had thought of my travelling impact as part of that, so [now] I'm like 'wow, my carbon footprint is probably absolutely enormous because of my travel'. I had never really thought about it that way.
>
> (ID 26)

The above traveller does not deny the impact of his extensive travels, he simply argues he had 'never really thought' about this impact. While two other travellers argued similarly, many others utilized a number of arguments to deny their responsibility, influence or impact. Denial strategies have been well documented in the environmental mitigation literature (e.g. Schahn 1993; Stoll-Kleemann et al. 2001; Lorenzoni et al. 2007). The interviewed travellers' arguments commonly related to Schahn's (1993) strategies of 'denial of responsibility', 'fabricated constraints', and 'comfort'. A total of 25 travellers considered their actions could have no impact (8) or little impact (17) on mitigating climate change. One interviewee (ID 8), for example, argued that everything people in 'Western Europe' do to mitigate climate change was futile due to the economic growth and subsequent increase in pollution of countries in 'Asia' (denial of responsibility). Furthermore, the perceived higher price and poorer quality of environmentally friendly products was noted by three travellers as a reason to refrain from buying them (fabricated constraints). Finally, 'comfort' was reported to be an issue, especially in the case of recycling and flying. Indeed, none of the respondents saw an alternative to travelling if they were to do their job properly.

To return to the quote above, when asked why he had not thought about his impact on the environment, the traveller was unable to give an answer. He did note, however, that at home he considered his family's environmental impact in many actions and decisions. This was a recurring phenomenon in interviews, with many travellers reporting themselves to be conscious about their environmental impact when in a home situation, but not so when travelling. This divide between home and 'away' behaviour has been reported by other scholars (Barr et al. 2010; Miao and Wei 2013). Travellers said that they became less environmentally friendly when travelling. This could be due to the unfamiliarity of the visited place, with some travellers arguing they refrained from recycling while travelling because there was an absence of recycling bins and their knowledge about customary procedures in visited countries was less than it was at home. Another traveller (ID 31) took taxis instead of public transport because of the convenience of getting to the hotel or being dropped at another destination without having to research what route to take. These examples portray how issues of unfamiliarity, convenience and time pressure can negatively influence travellers' environmental behaviour.

The fact that regular travellers often did not enjoy travelling, and experienced it as a mental and physical strain, arguably also influenced their perceptions about hotels, and consequently their behaviour while using them. The following quote from a Canadian traveller clarifies how his perception of a hotel stay is impacted by his busy working schedule: 'It's always really nice and relaxing to come up to a hotel room, like the bed is all made and you can just kind of fall down onto it and go to sleep' (ID 13). After a long day of work the hotel is a place where travellers can withdraw and relax. A clean hotel room (or the knowledge that the room will have been cleaned upon return) helps alleviate stress, as does knowing that hotel staff knows one's preference concerning pillows (as discussed above). This notion of 'knowing that it will be taken care of' is important to travellers because of their often busy schedules while travelling, and the toll this takes on their mental and physical fitness. It leads to many travellers expecting to have a 'frictionless' hotel stay, where human contact is not needed but 'available' when required or desired.

Interviewed hotel managers said that the implementation of environmental practices was based around this notion of 'friction'. The chief engineer of a five-star hotel situated in the London business district argued,

> I guess we [implement environmental practices in a discreet manner] because we don't want to, sort of, irritate the guest, or hassle the guest with any extra things, you know. It's just part of the whole concept of being a business hotel; you don't want to bother them.

This quote depicts a business hotel as a place that should be without 'hassle', allowing travellers to stay without having to deal with any hotel requests or information about environmental practices. Guests have certain expectations when coming to a hotel, and according to hotel managers these expectations do not always include being asked to consider their environmental impact. Other hotel managers discussed how hotel guests would often try to 'use as much stuff as possible' and were less 'conscientious as they are when they are at home'. These hotel managers, then, argued that business (as well as leisure) travellers have expectations of a hotel stay that do not necessarily include minimizing their environmental impact. Indeed, for none of the interviewed business travellers, were a hotel's environmental practices an important selection factor when booking. As one British consultant argued: 'Nobody is going to stop in a Marriott hotel because they bloody recycle their towels, or because they've got an environmentally friendly sticker in the bathrooms saying "we are environmentally friendly"' (ID 10). There are different aspects that contribute to the perceived unimportance of environmental practices – like reusing towels – among business travellers. First, other selection criteria were considered more important when selecting a hotel; price, location, brand name or standard of service. Second, reusing towels was argued by travellers to alleviate only a small part of the overall environmental impact of a hotel's operations, and hence could be seen as an 'insignificant' effort. A number of travellers viewed the towel reuse programme as a money-saving venture rather than a responsible business practice, with one traveller (ID 30) referring to the practice as a 'smarmy' effort by hotels to save money by disguising it as an effort to reduce the environmental impact of hotel operations. Indeed, almost half of the interviewed business travellers (15) noted the inherent wastefulness of hotel operations. The following quote from a female traveller explains how, despite the strains of travel, she consciously decides to participate in the towel reuse programme:

> I don't need my towels changed every day; I don't change them every day at home so I will leave it on the peg. In some hotels they still take it away

and give you fresh towels, and that happens on a daily basis. I think, 'what's the point of having that notice, and what's the point of me putting my towels on the peg'. So that's in many respects quite frustrating.... I can see the benefit of [complaining], but by the time I get back to the hotel and realise that they've done it, it's been a long day, I'm tired, and [I] just leave it.

(ID 6)

This traveller draws a comparison between staying in a hotel and being at home, arguing that she does not change her behaviour (the amount of towels used) when staying in a hotel, which goes against the argument of hotel managers discussed above. The quote furthermore illustrates how this traveller perceived hotels as wasteful places, where hotel staff would not always comply with her (pro-environmental) choices. The hotel thereby neglected to help her reduce the impact of her stay, and missed opportunities to prove to her that there was a genuine concern about their environmental impact. Finally, the traveller says she refrains from complaining because she has had a long day and is tired. Having to go through the necessary procedures to make a complaint would add 'friction' to the hotel stay, with the traveller seemingly unwilling to do this after a tiring working day. It could be argued, then, that regular travellers' experiences of travelling, and the mental and physical strains it imposes upon them, influences their perceptions of hotels and such establishment's attempts at pro-environmental behaviour.

Conclusion and implications

This study has sought to research how the lifestyles of business travellers impact their pro-environmental *in situ* behaviour. Through qualitative research which included business travellers and hotel managers, this study has made an attempt to consider factors that impact on travellers' willingness to participate in hotels' environmental practices, other than the availability of these practices. The mental and physical strain of travelling negatively impacted the attitudes towards travelling of the interviewed business travellers. This chapter has argued that this strain of travelling – as well as the negative attitude towards travelling overall – also negatively impacts the environmental considerations of travellers. With stressful and long working hours, in strange environments, most travellers expect hotels to be a place where they can withdraw and relax. Environmental considerations can often be impacted by this need for relaxation, but this chapter has also shown that many travellers are still willing to participate in environmental practices if they are easily accessible and if followed up by hotel staff. Future research is needed to further explore the motivations for travellers to partake in particular environmental practices.

The findings discussed in this chapter imply that the willingness of hotel guests to participate in environmental practices is not solely based on the availability of these practices or the motivations of guests, but also on other factors that influence their behaviour. This chapter has mainly focused on the strain of travelling as a factor influencing business travellers' behaviour, but future research is needed to establish whether and what factors further influence this. Future studies can establish how these inhibiting factors can be adapted to enhance environmental behaviour amongst travellers.

Activity

A question that runs through this chapter, but has not been answered here, is who is fundamentally responsible for the initiation and furthering of environmental practices and behaviour in the industries associated with business travel. Is it the responsibility of individual business travellers to change their behaviour, or should they be encouraged and helped by hotels or their employers? This is a question that for many decades has incited a debate among scholars writing on Corporate Social Responsibility.

This activity aims to allow students to investigate and discuss who should be held responsible for furthering environmentally and socially responsible behaviour. Divide the class in two groups. Every student should first read Friedman's (1970) influential article, followed by the article written by Kolstad (2007). Friedman argues that the only responsibility of corporations is to generate profit. Kolstad is more objective in his reasoning, but gives arguments to disprove Friedman. In their groups, students should answer the following questions:

- What arguments does Friedman use to reason that corporations' sole purpose is to make profit?
- What arguments does Kolstad use to disprove Friedman's reasoning?
- Can you think of recent developments in the world that can support or disprove the arguments of the authors?

One group will argue in favour of the social responsibility of companies, while the other group will argue against. Hold a structured debate about the key arguments of the two articles. If the groups consist of too many students, assign a number of students as representatives of each group. Reiterate to students that there is no right or wrong answer, and (where possible) let students argue in favour of the viewpoint they personally disagree with.

Further reading

There are two excellent books for an introduction on business travel:

Davidson, R. and Cope, B. (2003) *Business Travel: Conferences, Incentive Travel, Exhibitions, Corporate Hospitality And Corporate Travel*, Harlow, England: FT Prentice Hall.
Swarbrooke, J. and Horner, S. (2001) *Business Travel And Tourism*, Oxford: Butterworth Heinemann.

For interesting research on business travel see:

Lassen, C. (2006) 'Aeromobility And Work' in *Environment and Planning A*, 38 (2): 301–12.

Lassen further discusses the relationship between business travel and environmental considerations in:

Lassen, C. (2010) 'Environmentalist In Business Class: An Analysis Of Air Travel And Environmental Attitude' in *Transport Reviews* 30 (6): 733–51.

Notes

1 Airports consist of a 'landside' area, which is where travellers check in for their flights, and an 'airside' area, where travellers go after passing through security.
2 One of the main issues with interviewing travellers at London City Airport was the extensive use of laptops and mobile phones while waiting in the airport lounge. Instead of interacting with the researcher, or other travellers, most travellers exhibited a greater interest in interacting with their gadgets.

References

Barr, S., Shaw, G., Coles, T. and Prillwitz, J. (2010) '"A Holiday Is A Holiday": Practicing Sustainability, Home And Away' in *Journal of Transport Geography*, 18 (3): 474–81.
Becken, S. (2007) 'Tourists' Perception Of International Air Travel's Impact On The Global Climate And Potential Climate Change Policies' in *Journal of Sustainable Tourism*, 15 (4): 351–68.
Bell, D. (2007) 'The Hospitable City: Social Relations In Commercial Spaces' in *Progress in Human Geography*, 31 (1): 7–22.
Brotherton, B. and Wood, R.C. (2000) 'Hospitality And Hospitality Management' in C. Lashley and A.J. Morrison (eds), *In Search Of Hospitality: Theoretical Perspectives And Debates*, Oxford: Butterworth-Heinemann: 134–56.
Budeanu, A. (2007) 'Sustainable Tourist Behaviour – A Discussion Of Opportunities For Change' in *International Journal of Consumer Studies*, 31: 499–508.
Clausing, J. (2008) 'Boomers More Likely To Go Green In Business Travel' in *Travel Weekly*, 67 (2): 22.
Davidson, R. (1994) *Business Travel*, Harlow, England: Addison Wesley Longman Limited.
Davidson, R. and Cope, B. (2003) *Business Travel: Conferences, Incentive Travel, Exhibitions, Corporate Hospitality And Corporate Travel*, Harlow, England: FT Prentice Hall.
De Grazia, S. (1962) *Of Time, Work And Leisure*, New York: Twentieth Century Fund.
Douglas, A. and Lubbe, B.A. (2009) 'Violation Of The Corporate Travel Policy: An Exploration Of Underlying Value-Related Factors' in *Journal of Business Ethics*, 84 (1): 97–111.
Douglas, A. and Lubbe, B.A. (2010) 'An Empirical Investigation Into The Role Of Personal-Related Factors On Corporate Travel Policy Compliance' in *Journal of Business Ethics*, 92 (3): 451–61.
Elliott, A. and Urry, J. (2010) *Mobile Lives*, Abingdon, Oxon: Routledge.
Faulconbridge, J.R., Beaverstock, J.V., Derudder, B. and Witlox, F. (2009) 'Corporate Ecologies Of Business Travel In Professional Service Firms' in *European Urban and Regional Studies*, 16 (3): 295–308.
Friedman, M. (1970) 'The Social Responsibility Of Business Is To Increase Its Profits' in *New York Times Magazine*, 13 September.

Glaser, B.G. and Strauss, A.L. (1967) *The Discovery Of Grounded Theory: Strategies For Qualitative Research*, London: Weidenfeld and Nicolson.

Goldstein, N.J., Cialdini, R.B. and Griskevicius, V. (2008) 'A Room With A Viewpoint: Using Social Norms To Motivate Environmental Conservation In Hotels' in *Journal of Consumer Research*, 35: 472–82.

Greater London Authority (2002a) *Demand And Capacity For Hotels And Conference Centres In London*, London: City Hall.

Greater London Authority (2002b) *Visit London: The Mayor's Plan For Tourism In London*, London: Greater London Authority.

Gustafson, P. (2006) 'Work-Related Travel, Gender And Family Obligations' in *Work, Employment & Society*, 20 (3): 513–30.

Harris, C. and Ateljevic, I. (2003) 'Perpetuating The Male Gaze As The Norm: Challenges For "Her" Participation In Business Travel' in *Tourism Recreation Research*, 28 (2): 21–30.

Harris, C. and Wilson, E. (2007) 'Travelling Beyond The Boundaries Of Constraint: Women, Travel And Empowerment' in A. Pritchard, N. Morgan, I. Ateljevic, and C. Harris (eds), *Tourism & Gender: Embodiment, Sensuality And Experience*, Wallingford, Oxfordshire: CAB International: 233–50.

Hawkins, R. (2006) 'Accounting For The Environment: Reflecting Environmental Information In Accounting Systems' in P. Harris and M. Mongiello (eds), *Accounting And Financial Management – Developments In The International Hospitality Industry*, Oxford: Butterworth-Heinemann: 262–81.

Ivancevic, J.M., Konopaske, R. and DeFrank, R.S. (2003) 'Business Travel Stress: A Model, Propositions And Managerial Implications' in *Work and Stress*, 17 (2): 138–57.

Kang, K.H., Stein, L., Heo, C.Y. and Lee, S. (2012) 'Consumers' Willingness To Pay For Green Initiatives Of The Hotel Industry' in *International Journal of Hospitality Management*, 31 (2): 564–72.

Kirk, D. (1995) 'Environmental Management In Hotels' in *International Journal of Contemporary Hospitality Management*, 7 (6): 3–8.

Kolstad, I. (2007) 'Why Firms Should Not Always Maximize Profits' in *Journal of Business Ethics*, 76: 137–45.

Lashley, C. (2000) 'Towards A Theoretical Understanding' in C. Lashley and A.J. Morrison (eds), *In Search Of Hospitality: Theoretical Perspectives And Debates*, Oxford: Butterworth-Heinemann: 1–17.

Lassen, C. (2006) 'Aeromobility And Work' in *Environment and Planning A*, 38 (2): 301–12.

Lassen, C. (2009) 'A Life In Corridors: Social Perspectives On Aeromobility And Work In Knowledge Organizations' in S. Cwerner, S. Kesselring, and J. Urry (eds), *Aeromobilities*, Abingdon: Routledge: 177–93.

Lassen, C. (2010) 'Environmentalist In Business Class: An Analysis Of Air Travel And Environmental Attitude' in *Transport Reviews*, 30 (6): 733–51.

Lockwood, A. and Jones, P. (2000) 'Managing Hospitality Operations' in C. Lashley and A.J. Morrison (eds), *In Search Of Hospitality: Theoretical Perspectives And Debates*, Oxford: Butterworth-Heinemann, 157–76.

Lorenzoni, I., Nicholson-Cole, S. and Whitmarsh, L. (2007) 'Barriers Perceived To Engaging With Climate Change Among The UK Public And Their Policy Implementations' in *Global Environmental Change*, 17 (3–4): 445–9.

McNeill, D. (2008) 'The Hotel And The City' in *Progress in Human Geography*, 32 (3): 383–98.

Miao, L. and Wei, W. (2013) 'Consumers' Pro-Environmental Behaviour And The Underlying Motivations: A Comparison Between Household And Hotel Settings' in *International Journal of Hospitality Management*, 32: 102–12.

Millar, M. and Baloglu, S. (2011) 'Hotel Guests' Preferences For Green Guest Room Attributes' in *Cornell Hospitality Quarterly*, 52 (3): 302–11.

ONS (2011) *International Passenger Survey*. Online, available at: www.statistics.gov.uk/ssd/surveys/international_passenger_survey.asp (accessed 30 April 2011).

Pritchard, A. and Morgan, N. (2006) 'Hotel Babylon? Exploring Hotels As Liminal Sites Of Transition And Transgression' in *Tourism Management*, 27 (5): 762–72.

Ritzer, G. (2004) *The McDonaldization Of Society – Revised New Century Edition*, London: Sage Publications Ltd.

Ritzer, G. (2007) 'Inhospitable Hospitality?' in C. Lashley, P. Lynch, and A. Morrison (eds), *Hospitality: A Social Lens*, Oxford: Elsevier: 129–39.

Schahn, J. (1993) 'Die Rolle von Entschuldigungen und Rechtfertigungen für umweltschädigendes Verhalten' [The Role Of Excuses And Justifications In Environmentally Damaging Behaviour] in J. Schahn and T. Giesinger (eds), *Psychologie für den Umweltschutz* [The Psychology Of Environmental Protection], Weinheim: Beltz: 51–61.

Schmidt, C. (2004) 'The Analysis Of Semi-Structured Interviews' in U. Flick, E. Von Kardorff and I. Steinke (eds), *A Companion To Qualitative Research*, London: Sage Publications: 253–8.

Sheller, M. and Urry, J. (2006) 'The New Mobilities Paradigm' in *Environment and Planning A*, 38: 207–26.

Solomon, J. (2009) 'Happiness And The Consumption Of Mobility' in K. Soper, M. Ryle, and L. Thomas (eds), *The Politics And Pleasures Of Consuming Differently*, Basingstoke: Palgrave Macmillan: 157–70.

Stoll-Kleemann, S., O'Riordan, T. and Jaeger, C.C. (2001) 'The Psychology Of Denial Concerning Climate Mitigation Measures: Evidence From Swiss Focus Groups' in *Global Environmental Change*, 11(2): 107–17.

Sunstein, B.S. and Chiseri-Strater, E. (2002) *Field Working: Reading And Writing Research*, 2nd edition, New York: St. Martin's.

Swarbrooke, J. and Horner, S. (2001) *Business Travel And Tourism*, Oxford: Butterworth Heinemann.

The Independent (2011) 'London Tops Ranking Of Destination Cities'. Online, available at: www.independent.co.uk/travel/news-and-advice/london-tops-ranking-of-destination-cities-2291794.html (accessed 5 April 2013).

Tierney, R. (2007) 'Going Green: Sustainable Practices Take Root In Hospitality' in *HSMAI Marketing Review*, Summer 2007: 24–33.

Tzschentke, N., Kirk, D. and Lynch, P.A. (2004) 'Reasons For Going Green In Serviced Accommodation Establishments' in *International Journal of Contemporary Hospitality Management*, 16 (2): 116–24.

Urry, J. (2007) *Mobilities*, Cambridge: Polity Press.

Urry, J. (2009) 'Aeromobilities And The Global' in S. Cwerner, S. Kesselring and J. Urry (eds), *Aeromobilities*, Abingdon: Routledge: 25–38.

13 Medical tourism

Consumptive practice, ethics and healthcare – the importance of subjective proximity

Kirsten Lovelock and Brent Lovelock

Introduction

People travelling across and within national borders to procure a range of healthcare services is an increasingly prevalent phenomenon. Driven by a range of forces, and provoking the rapid development of specialised health services in destination countries and regions, medical tourism involves a wide range of countries and regions globally. Defined variously, medical tourism involves the intentional pursuit of non-emergency medical treatment outside one's home country (Crooks *et al.* 2010), can involve travel within one's own country to another health care jurisdiction (domestic medical tourism), and represents an individual solution to a health system problem that has historically been addressed by the citizen's governmental body (Pocock and Phua 2011). Importantly, unlike state-funded cross border treatment, the medical tourist initiates and covers all treatment expenses.

There are a range of ethical issues central to the commodification and trade of healthcare and the consumptive practice of medical tourism. Some of these issues include; the effect of medical tourism on health service provision in both provider and departure societies; whether medical tourism contributes to the development and/or perpetuation of two-tiered health systems where the wealthy receive quality private care and the poor continue to struggle to receive basic public healthcare; the role played by medical tourists in the spread of infectious diseases and antibiotic resistance; risks for patients; and emergent questions over who is responsible for the long-term care of medical tourists when treatment abroad fails (Lovelock and Lovelock 2013). While the ethics of medical care has a long history, it has typically been addressed within borders and national healthcare systems. Medical tourism presents additional ethical challenges. These challenges are provoked by the globalised nature of the trade in healthcare and services, the involvement of a diverse range of stakeholders (not all of whom are health professionals) but who are driven by diverse motivations, to facilitate non-essential and essential treatment abroad.

This chapter reports on a content analysis of medical tourist experiences posted on the internet with the intention of exploring emergent issues surrounding the ethics of globalised healthcare consumption. We focus on individual medical tourists who have sought essential treatment abroad, rather than those whose insurance companies or states have provided incentives for public employees, or citizens, to seek treatment abroad for non-essential treatment (Connell 2011, 2012). The blogs on these internet sites provide an insight into the motivations for seeking medical treatment abroad, allow us to explore the ethical implications, and suggest that consequentialist frameworks, favoured by social scientists when addressing ethical consumption, are unlikely to be embraced by medical tourists seeking *essential* medical treatment.

The emergence and growth of medical tourism

There are a range of drivers underpinning the emergence of medical tourism. In high income developed countries, ageing populations and the burden of increasing rates of chronic disease are straining health systems and contributing to what are increasingly thought to be unsustainable healthcare costs. Medical tourists are motivated to travel away from home for treatment for a range of reasons, including: long waiting lists for health services and treatment; prohibitively expensive treatment costs for some services; and the non-provision of some services – either because of cost or because of bio-ethical legislation (for example: abortion; assisted conception). Medical tourism is a distinct niche within health tourism, and is distinguished from other forms of health travel because it involves medical intervention (Connell 2006; Balaban and Marano 2010; Connell 2011; Hall 2011). In addition, with medical tourism, the medical intervention and health service is marketed alongside the touristic potential of taking the trip and the post-operative bonus of recovering in an exotic location (Connell 2011; Lovelock and Lovelock 2013).

In the last two decades medical tourism has emerged as a multi-billion dollar industry, one that predominantly involves people from high-income countries seeking treatment in low-income countries (Crooks *et al.* 2010) and where most of the medical tourists come from North America, Western Europe and the Middle East and South Asia. In addition, particular regions have emerged as specific treatment destinations, with patterns of treatment emerging where medical tourists show preferences for certain destinations depending on their ethnicity, religious observances, proximity and cost (Connell 2006). Thus, Middle Eastern medical tourists tend to seek treatment in Malaysia, Omanis in India, and Japanese in Thailand and Singapore. Destinations have also specialised, and increasingly market themselves according to specific medical interventions, becoming a (the) place for cosmetic surgery, dental work, transplants and heart surgery, transplant surgery, experimental treatments etc. (Connell 2011).

There are a range of ethical issues provoked by medical tourism, and central to these issues is the debate surrounding the neoliberal informed liberalisation of health services and the erosion of public health service provision in both destination and departing countries. One of the primary ethical concerns is that medical tourism will impact negatively on healthcare provision and access in the poorer destination countries. The processes of neoliberal reform in advanced capitalist societies have ensured that health services have become a tradeable commodity (Hermans 2000). This trade involves foreign investment in healthcare, a global network of health insurers, telemedicine, and with medical tourism a range of other service suppliers that facilitate the movement of people across borders for the consumption of medical treatment and immediate (if not long-term) recovery (Pennings 2007; Cortez 2008; Lovelock and Lovelock 2013). Healthcare has emerged as one of the world's most significant markets, and the internet is central to engagement with this market, providing information on health conditions, providers of treatment, and links to brokers who can facilitate the medical tourist's itinerary (Lunt *et al.* 2010).

The internet is now a major facilitator and shaper of healthcare consumption provoking a number of issues with respect to the quality of information and advice provided to patient consumers. Lunt *et al.* (2010: 7) note it is rare to encounter an internet site that is non-commercial in nature. The emergence of web-based advertising by various health centres (health experts within a commercial enterprise), and where patient satisfaction (personal experience) is used to attract paying patient self-referral (Lunt and Carrerra 2011), demonstrates clearly that healthcare has become a tradeable commodity and a form of consumption which is often initiated by the individual patient. Health sites provoke concerns about the quality of information provided, but they also provide an important research vehicle for understanding what issues are central for those seeking treatment and how these businesses capture health consumers by explicitly addressing these issues.

The movement of people seeking treatment across national borders has led to the suggestion that we live in the age of the 'global biological citizen'. What ethical issues are provoked by these global citizens and this increasingly dispersed and complex industry? To date most of the research on the ethics of medical tourism has focused on consumer rights and safety, and, ultimately in societies where medical care is not always assured to be provided by the state, issues surrounding liability and the inability to address malpractice (Burkett 2007). However, a broader view is necessary and one that embraces the fact that medical tourism – or the therapeutic itineraries typical of medical tourists – are itineraries that traverse significant gradients of socio-economic and political inequality and indeed may significantly contribute to its reproduction (Bose 2005; Connell 2011; Snyder 2012; Lovelock and Lovelock 2013). This broader view allows consideration of the ethical implications of the impact medical tourism has on healthcare provision in destination regions; the potential diversion of interest from domestic patients to higher-paying foreign medical tourists; continued neglect and/or an abdication of responsibility towards those experiencing inequitable access to healthcare in their home countries; the commodification of global biological resources; the legitimacy of transplant sources; and the ethics of health consumption and the commodification of healthcare where some have limited purchasing power and differential access to purchasing options.

The ethics of consumption: consuming travel for medical intervention – or consuming medical intervention through travel

Medical tourism immediately provokes the taken-for-granted assumption that space and place are factors that can inhibit or facilitate moral duties, and where it might be assumed that distance is a major facilitator of indifference (Smith 2000; Barnett *et al.* 2005). The commodification of healthcare is evident with medical tourism, where travel connects illness to consumer choice and provokes hope about circumventing failing public health systems at home (Solomon 2012). Travelling for healthcare consumption offers the possibility to explore political (where the political is personal) and ethical responsibility, and to raise a number of questions. For example: what can the role of medical tourists as consumers in this consumption process tell us about political and ethical responsibility in relation to healthcare? Does spatial distance lead to indifference, loss of social responsibility and less care for others? Or does this form of consumption allow for ethical transformation and a modified form of political agency? And, if the political is personal – what role does subjectivity play in medical travel?

How might we conceptualise the ethics of consumption in relation to medical tourism? Is spatial scope important? Does having treatment at a 'distance' provoke ethical issues that are specific *to* distance? Does knowledge of a destination lead to responsible action? Responsible and moral action is often assumed to be an outcome of knowing that actions have consequences and knowing what actions in what contexts will provoke what outcomes. With respect to consumption, this is just one point in a linear chain which commences with production and distribution, and ends with consumption. This chain comprises differing commitments and responsibilities, and ultimately consequences (Barnett *et al.* 2005), and it is assumed in the social sciences that when these processes are exposed and come to be known (cease to be taken for granted) moral and responsible action can occur or is facilitated.

This is a consequentialist model of ethics and it is this model that is dominant in debates concerned with ethical consumption. This model is, as we will see, problematic, as at its core it is individualistic and assumes that moral conduct (moral agency) must be underpinned by knowledge. The model does not allow for social relationships which support or impede moral conduct (Barnett *et al.* 2005) and assumes that humans are reflective subjects who will change behaviour once exposed to information or knowledge. In most areas addressing ethical consumption – from fair trade to sustainable consumption to corporate responsibility, it is assumed that ethical conduct is dependent on knowledge and information – that is, when people know and are informed they will make an explicit commitment to behavioural change (Barnett *et al.* 2005).[1] Yet we know from the tourism literature which addresses ethical consumption that people have a habit of not behaving in such a straightforward way (for example, pro-poor tourism, eco-tourism). For the purposes of this chapter we will be employing a perspective taken from a range of research that emphasises the importance of routine (or in our words 'habitual') consumption and learned ethical competencies (Hobson 2002; Barnett *et al.* 2005). We will consider how consumption occurs in relation to 'others', not simply as an individualistic self-indulgent pursuit (Barnett *et al.* 2005: 28). The key question underpinning this chapter is: what counts as ethical consumption in the field of medical tourism – and from whose perspective?

Ethical consumption can be conceptualised as comprising two dimensions: (1) an organisational dimension that facilitates the 'ought to' into 'can do', and; (2) an inter-subjective dimension which involves how individuals govern their consumption (Barnett *et al.* 2005). This governance is shaped by how individuals frame the self–other relationship (which is infinitely variable) and also how practices demonstrate specific ethical competencies (ibid.). Here the person does not take an abstract model and apply it to a situation, rather they explore how various practices enable them to articulate their core moral and ethical competencies. This way of conceptualising ethical consumption is very useful when considering medical tourism from the medical tourist's perspective.

We know little of the personal experience of medical travel, and in particular how the ethics of this practice are perceived and negotiated by medical tourists as consumers of healthcare. While we know that medical tourism has transformed the geography of healthcare, we know little of the role that websites play, or of the role significant others play in medical tourists' lives, in terms of their motivations and decision-making processes; we also know little of the behaviour of medical tourists once at their destination (Connell 2012). Crooks and colleagues' (2010) scoping review of patients' experiences of medical tourism found people usually addressed three things; (1) positive and negative aspects of medical tourism; (2) sensationalised issues, and; (3) reports of post-recovery life. Pre-travel assumptions were often challenged on arrival, and many of the assumptions hinged on concerns about competency and hygiene (Crooks *et al.* 2010); and these concerns were often shared by family members and friends following the decision to seek treatment abroad. As a component of the same study, Crooks *et al.* (2012) also conducted semi-structured interviews (n=32) with Canadian medical tourists, focusing on the decision-making process. They found that the cost and availability of procedures were primary motivators, and that the internet was the primary facilitator for establishing connections with overseas providers. Importantly too, the experiences of other medical tourists shaped their decision-making. The researchers concluded that the decision-making process was multi-layered and not driven by any single primary motivator. Snyder *et al.* (2012) have developed an ethical decision-making model that potential medical tourists can follow when considering seeking treatment abroad. This is a consequentialist model and provokes a number of questions. For example, would a person contemplating travelling abroad for medical treatment, given knowledge about potential inequities that they may contribute to the perpetuation of, change their mind about seeking treatment? We would suggest, drawing on our reading of blogs on a medical tourism site, that for most of these people this would be unlikely, primarily because their ethical decision-making is shaped by *habitual ethical frames*, and travelling for treatment within these frames constitutes 'ethical conduct'. As we will see, what we call '*subjective proximity*' is central to seeking treatment abroad, and *subjective proximity* has replaced *geographic proximity* as a guarantee for access to healthcare. Further, we argue, the commodification of illness has ensured the 'real' measure of distance (distance that matters) is subjective.

Method

Our method involved searching the internet using the Google search engine, and searching with the following keywords and combinations of these words: 'medical', 'tourism', 'health', 'travel', 'treatment', 'abroad', 'private', 'healthcare'. One dominant site was identified, and 35 blogs on this site were downloaded. Only those travelling for essential health treatment were included (n=35), and the dominant treatments were heart surgery and hip resurfacing, with the majority having treatment in India. All of the blogs are retrospective accounts of patients' experiences of medical interventions carried out abroad. All of the blogs are by North Americans, with the majority from the United States of America. The blogs were printed out, the text coded, and themes identified. The themes were then subjected to interpretative content analysis (Krippendorff 2004; Smith 2010).

Smuddle (2005) has identified four categories of blog and for the purposes of this chapter we are focusing on the first category – personal blogs – where people express their personal convictions, observations and suggestions; and in this instance they do so through a 'commercial' site, where their statements also stand as testimonials. We are aware it is possible that personal blogs that are critically reflective of the process might not be posted on such a site and we are therefore unable to address this experiential data. We are not making any claim to representativeness or generalisability, and fully acknowledge the limitations of web-based research with respect to reliability and contextual deficiency (Banyai and Glover 2011). Rather, our intention is to explore some emergent themes surrounding the ethics and conceptualisation of the consumption of medical treatment overseas. We also think it is important to address these blogs as they contain the dominant themes used to promote and attract this category of health consumer in what is clearly an increasingly important and financially successful industry.

Context of treatment

Many of the bloggers on this site had sought hip resurfacing in India and, in few numbers, heart surgery in the same destination and mainly from the same provider – Apollo Hospital Group (AHG) – which has accreditation from Joint Commission International (JCI). This hospital group had treated tens of thousands of medical tourists seeking heart surgery and joint surgery, among a range of other procedures and interventions on offer. This group has facilities in Delhi,

Chennai and Hyderabad. Many users reported disappointment with their experiences of healthcare services in their home state, in the United States of America, or province, in Canada. Hardly any could afford to cover the cost of their treatment at home, and were either uninsured or their insurance company would not, for a variety of reasons, cover the cost of treatment. These medical tourists represent some of the 46 million Americans that currently do not have health insurance coverage (Meghani 2011) and the growing number who seek treatment abroad. The Patient Protection and Affordable Care Act and the Health Care and Education Reconciliation Act of 2010 represent a move to address this significant health inequality in the United States, but even after these Acts have been fully enacted approximately 23 million Americans will remain uninsured (Meghani 2011). The vulnerable uninsured are: the poor, the unemployed, low-income earners, the self-employed and employees of small businesses. The neoliberal underpinnings of this situation, where insurance companies can set the premiums and where these premiums have become prohibitive for many, underpins the significance of medical tourism in providing essential treatments to North American citizens – not to mention state abdication of public health responsibility.

The postings

None of the bloggers overtly reflected on the universalistic ethics of medical tourism and the implications for either their own, or others', healthcare, services or systems. Yet this does not necessarily mean that ethics or understandings of moral conduct do not underpin their motivations, decision-making and behaviour. The key emergent themes are discussed below.

Consume: do or die

For those who had gone to India for heart surgery the choice of staying at home or going abroad for treatment was a choice of *doing something* or *staying home to die*. In the words of one wife, describing her husband's situation in the United States, 'They sent him home to die.' And, in the words of another, describing his health situation,

> My surgeon told me 'Well I can keep you alive for about 3–5 years on medications' and so they stabilised me and sent me home. They quoted $200,000 for a triple bypass ... and I said 'well I am not going to India' [a statement he later revisited].

We know that cost and long waiting lists are major motivators for those who travel abroad for a range of medical treatments and interventions, but there is little research that addresses the subjective dimensions of what the 'cost' and 'wait' mean for those who are seeking essential treatment. We think Barnett *et al.*'s (2005) concept of 'moral-selving' is useful for understanding how routinised and habitual behaviours facilitate the governance of self. The blogs on this site demonstrate that medical tourism offers these people hope, and it is also clear that for these medical tourists their ethical decision-making is couched in terms of their relationships with family – husbands, wives, children – and/or close friends, and their sense of moral and ethical obligation towards these significant people. Because of the nature of seeking treatment (to stay alive or die – or for those seeking hip resurfacing, staying in a state of perpetual pain and being a burden on those who care for them) it is perhaps not surprising that few explore 'going-beyond-the-self, in a deliberate attempt to achieve degrees of selflessness in order to practice responsibilities to distant others' (Barnett *et al.* 2005: 31). The particular is their focus, not the universal. The reason is quite simple; their concern is not with distant (unrelated in the broadest sense) others, but with *those who they immediately care for – and who care for them –* what we call *subjective proximity* – and this is the key characteristic of *their ethical focus*. Ultimately, subjective proximity is expanded to include those in other places – and replaces the historically based assumption that geographic proximity will ensure access to healthcare and, more generically, 'care' (this is discussed more fully in Lovelock and Lovelock, forthcoming).

The 'ought to' for the bypass patients is an 'ought to stay alive' and to seek the most financially viable treatment. The organisational dimensions comprise the medical tourism industry; the 'moral selving' is shaped by the self–other relationship of kinship (either actual or proxy); and the obligation to stay alive is underpinned by the need to ensure that the survival of the family or significant relationships are not undermined by the illness. For these medical tourists who are seeking essential medical treatment, the moral and ethical positioning has been individualised and thus it necessarily becomes micro level, and bypasses society or broader social processes. This is a direct outcome of their own society (governance) relinquishing responsibility for their health care. The commodification of healthcare, and the emergent system's dependence on the ability of subjects to purchase and pay, has ensured that the organisational dimensions require the uninsured person to either seek alternatives or continue to live in pain and place an unfair burden on their carers (family and proxy kin), or die. Making sense of seeking these alternatives, which place an unfair burden on those they care for, requires these medical tourists to employ habitual ethical and moral frameworks and to understand how their 'illness' relates to the new circumstances they find themselves in with respect to healthcare, and those who 'immediately' (time rather than spatially) can care for them.

Time

Delays or waiting lists:

> Unfortunately, getting medical procedures that don't involve life or death, like hip replacement, can be a painfully slow process.
>
> (Canadian, male, hip resurfacing)

Quick service – care and service abroad:

> Moe [the service provider intermediary] always called me back immediately (within minutes) on receiving my emails, day or night, weekends ... closer attention and better service than I receive in our hospitals in Canada'.
>
> (ibid.)

More generally, the response to seek treatment is also in part an outcome of what others have described as the 'time famine' plague in the United States [and arguably most late-developing capitalist societies], a famine (sic) which underpins Americans doing 'whatever it takes' to maintain a quality of life. The 'whatever it takes' is usually self-medicating, as they do not have time to be sick (Vuckovic 1999). However, it might equally apply to medical tourism because *life-time* is a scarce resource and the 'whatever it takes' becomes the consumption of health treatment abroad – medical tourism itself. Treatment abroad is buying time, and it is this buying of time that brings with it the chance of a better quality of life than heavy medication (or death, even) can offer. Furthermore, the choice to go abroad for treatment is a self-care choice that parallels already-established cultural norms of self-care – purchase and consume the remedy and treat your time as a scarce resource which should be utilised optimally (Zerubavel 1981, cited in Vuckovic 1999).

Waiting for treatment is antithetical to culturally entrenched understandings of 'time' – bearing a relation to social status, economic wellbeing and success – where there is no time to be sick and those who address the time deficiency brought on by sickness are behaving morally and optimally.

But perhaps most persuasively, this healthcare-seeking, for those who are uninsured, is an expression of affective agency toward an 'other' who responds affectively. The 'other' health provider (and in particular their intermediary) responds appropriately (morally) in the face of the uninsured person's distress. Ironically, the cultural other complies with their culturally prescribed way of communicating distress (Nichter 2010), their culturally entrenched means of managing poor health experience and its expression, and provides what was 'once' a culturally entrenched means of response (i.e. access to health services and healthcare).

Pain and disability

Ongoing pain and increased disability was another emergent theme and in the majority of instances was linked to the need for surgery to joints. For those who are in pain and experiencing increasing immobility, medical tourism is an option open to them which allows relief of pain and mobility, and as 'consumers' they seek the goods and services that are most effective (long term) and that are available at affordable prices. Their moral choices are framed by neoliberal rhetoric and realities – and they are compelled to embrace both rhetoric and realities because they, as 'uninsured others', largely accept they cannot change 'the system' which they have been excluded from. This is born out in the statements made about their engagement with insurance companies, as we will see below.

Insurance – against what?

> [T]he other big problem here [in the United States] is fighting with the insurance company to actually get them to cover what they say they will cover. And needing heart surgery you don't need the stress of having to fight with the insurance company to get what you paid for.
>
> (patient's wife, American, Colorado, heart surgery)

> I'm a middle class guy. I work for a living. I pay taxes, believe me ... it was pretty close to $700 a month in 2000 when my insurance company cancelled me.... The major stumbling block to me getting health insurance is the fact that I'm diabetic.... I knew there was something wrong but with no health insurance it's expensive even diagnosing the problem.... India seemed to be like the most bang for the buck.
>
> (American, Arizona, triple bypass surgery)

The central themes to emerge in relation to insurance companies were that *the unwell do not have the energy to fight* (insurance companies) and (ironically) *being unwell is a disadvantage when it comes to the ability to consume health care*. The two are related and provoke a stasis in a context where only the 'wider marketplace' can facilitate agency.

Choosing to travel for medical intervention is the choice to empower oneself over one's health and longevity; the power to seek essential treatment. While web-based medical travel sites promote medical travel as a socially acceptable option (Sobo *et al.* 2011), the promotion is only successful because it conforms to the dominant rhetoric of patient-as-consumer in a world where healthcare has been commodified and where there is a moral obligation to 'take care of your health' and to at least try to live a long and healthy life. Hence, people 'fight disease' and 'put up a fight' – they do not succumb willingly to death, and ironically they are failing if they allow 'ill health to consume' them. Such is the power of individual agency in the face of societal dissolution of taxpayer power.

For some, they are doubly impeded when it comes to dealing with insurance companies. With long histories of poor medical care and emergent chronic conditions, medical insurance companies will not cover them, or will not provide cover for particular conditions. That these people are unaware of a developing condition becomes, in the eyes of the insurer, evidence that they have not been prudent in their healthcare. As one woman explains below:

> Four days before the surgery [hip replacement] was scheduled my insurance company 'unauthorised' it ... they would not cover it. They said there had been a mistake. I pursued an informal appeal process ... [but] they

determined that a prudent person would have known they had osteo arthritis, though I had not seen a doctor, or had an X-ray, or had any previous medical history.... [E]ncouraged by my daughter ... she convinced me I needed to look for my medical care out of the country, where I could afford it.

What neoliberalism took away, the market can provide

The 'other'

All of the postings posit the self–other relationship as one where the self is firmly positioned as the sick/in pain/disabled self in relation to the 'other', who is able to relieve their suffering and/or those that assist them to get to 'the other' healthcare provider. The contrast between their developed society and the developing society of the 'other' is addressed in relation to health-related concerns: cleanliness and post-operative care (often accompanied with an expressed surprise at the level of care), and cleanliness and technological sophistication that 'the other' (being from a disadvantaged/poorer nation) can offer. None of the blogs reflect on what the implications of medical tourism might be for healthcare systems in the destination country, and none of the postings suggest a perception that they as medical tourist are comparatively 'the haves' rather than, 'the have-nots'. Rather, the emphasis placed by these medical tourists is on being the 'have-not's' when at home. Their frame of reference is local and when abroad, subjective – where subjective proximity obscures geographic distance and healthcare inequity in the destination country.

Intermediaries and facilitators

Many of those posting on the site spoke of the efficiency, kindness and care shown by those who serve as intermediaries or facilitators of the travel for treatment abroad. A man named 'Moe' was cited by a number of the medical tourists for his ability to listen, respond quickly, and to care, and for his knowledge on what could or couldn't be done. The post below was typical:

> Special mention should be made of Moe, the gentleman at Healthbase who works directly with patients. Moe helped with all the logistical aspects of the trip.... Moe was always available.... The most important aspect of Moe's behaviour was his positive outlook. I really cannot imagine a more reassuring individual with whom to work. He was always enthusiastic about [the] success of surgeries.

Attentive care

> He was easy to discuss my story with, his understanding manner.... [T]he hospital staff were excellent care givers.
> (Male, American, bilateral hip replacement)

Others noted that this level of attentiveness did not exist in the healthcare system at home. The intermediary, in this case Moe, but it might be said of others, becomes everything that their health system at home has lost – a 'face'/'voice' that represents care and the ability to respond to those who need care or treatment. Thus, Moe and other intermediaries, in this worldview, become almost proxy kin (after all they are the ones that answer a call – not an automated message) – they care as much as the unwell person's kin (or proxy kin) do, they listen, take them seriously and do all they can to assist in the procurement of treatment. These intermediaries are of course crucial to the business of medical tourism – but they are equally crucial to the medical tourist who seeks to be cured. Intermediaries are one step removed from the physician and are, *if not more importantly*, one step closer to the patient than the local health system is.

For many, distance is no longer measured in kilometres, rather it is measured subjectively. For many of these people a plane ride to another country makes sense subjectively when compared to a short trip across town to a medical facility they cannot afford to enter and where they cannot afford to procure care, treatment or advice; and where they know they will be treated with indifference. Distance does matter (Smith 2000; Barnett 2005), but it is not always the greater physical distance that matters; rather 'distance' from care, from treatment, from service, can also be measured socially and experientially (subjectively). For those who travel abroad for essential health treatment the social distance at home is what they cannot traverse; indifference in this context is local and relative to social status. They understand that healthcare is no longer locked into geographic proximity (because they have tried); they also understand that despite geographic distance and cultural difference care is being offered elsewhere and they can access this care and will not be treated indifferently. Despite the 'consumptive' nature of this experience (where a price is paid to receive the experience), health is not a 'product', and habitual ethical frames, in this respect, have remained historically constant; that is, when ill the subjective price to pay is too high and this subjective position drives the behavioural response.

A tourism experience?

> The clinic was like a five-star mini-hotel with a choice of Indian or Western food. Most of the patients were Americans....I was totally happy with the care.
>
> (Female, American hip resurfacing)

> After being here for over two weeks, my ear has developed more and I can understand the East Indian accent a little bit more. I spent little time vacationing in India. Most of my time was spent getting into the hospital, being here and recuperating, yes we did go to the zoo and to the Bay of Bengal. I would possibly consider coming back to India for treatment or vacation some time. Or I might consider another country too....People tell me 'Gee, you are really brave going to India for medical procedures!' For me it really

wasn't that much about being brave because I trust the universe. It was really fun being here. The people are wonderful. I love the fact that they don't slaughter cows and I have much respect for the Hindu religion. It was very interesting being in a totally different country.

(Female, American, uninsured, hip resurfacing)

Most of the blogs do not address the touristic aspect of the trip and treatment/recovery process in a stereotypical way. Other researchers have noted that those seeking challenging procedures or medical care because they cannot access it at home are seldom 'tourists' in the sense that they have chosen to go abroad for pleasure or for a vacation (Connell 2013: 3). Indeed the 'tourism' component is more commonly linked to those seeking elective treatments such as cosmetic surgery. Nonetheless, all postings address the travel aspects of seeking medical care and *subjective proximity* ultimately mediates this cognitive and behavioural shift.

In addition to this, the nature of treatment being sought largely predicates the extent of 'touring' that happens while they are abroad. Most of the patients commented on the food they consumed, the care received, and the nature of their surroundings. Arguably they would probably do the same if the treatment was being provided at home. However, for some, how they report these features is shaped by being abroad. As the quote above illustrates, there is an element of adventure, of discovery, and of the bravery that is assumed to go with those who travel abroad to 'exotic' locations; and as travellers and patients the journey involves encountering challenges and risks.

Encountering and managing these risks and challenges does provide 'tourist or traveller' kudos, and this kudos stands alongside the other moral achievements of procuring the cheapest deal, taking care of one's health, addressing disability and individually addressing what the system cannot provide. In many respects medical tourism offers a vehicle for social mobility in as much as it enables those who are not enabled at home; individuals, irrespective of social place, can lay claim to experiences that are common to the (insured) middle and upper middle classes. Medical tourists can experience the empowerment that comes with being able to afford travel, treatment, comfort and care, and when they return they can recall and retell their experience of the foreign country that provided that treatment and care while simultaneously having benefited from improved health and social standing.

Conclusion

We should be wary of completely buying into such particularism and relativism, of course; but it remains that such an analysis allows us to consider, more critically, ethical decision-making in relation to medical tourism. Medical tourists seeking essential treatment, including those who have blogged on the site we accessed, in many respects share the same concerns as those domiciled in those same host countries that are providing care for foreigners, but their circumstances of course differ. The extremes of poverty in the destination countries (such as India) are greater than those in North America (at least for some); many among the local population die from treatable diseases and conditions, and are not afforded the effective care afforded to the majority living in advanced capitalist nations. There remain stark disparities in India between the healthcare provided to locals and that provided to foreign fee-paying medical tourists. And while the impact medical tourism has on these disparities is still to be quantitatively documented (Connell 2011, 2012) the practice continues to provoke ethical concerns.

Key amongst these are, if the impact on healthcare in destination countries is adverse and undermines access to care for locals, then what are the home country obligations with respect to mitigating these negative impacts? Who benefits from medical tourism in the referring countries? And, who would care for the medical tourists if they were no longer able to travel for essential treatment? Most compellingly, it remains the case that for these medical tourists, seeking care in India is an individual solution to a local health system problem that was historically addressed by that citizen's own government. There may well be risks associated with medical tourism for both the individual and a collective public health establishment which is already under pressure from the neoliberal agency (Hall 2011), but the question is: what is the panacea for this? And, does this panacea rest with those whose illness has been 'commodified' – where their dollar for their 'illness' only has meaning abroad? The inability to provide adequate access to healthcare for those who need essential treatment, for whatever reason, feeds medical tourism while simultaneously feeding an ethic of 'self' care that is predicated on a lack of care to the 'citizen' public – ironically because those societies that tax on the basis of a 'citizen' public, no longer seem to 'care'.

It seems unlikely, however, that the answers to these ethical dilemmas lie in seeking (or demanding) a consequentialist perspective in order to realise social justice in this area, despite the obvious attraction. Even if it is known that there are negative impacts, the medical tourist seeking essential treatment will weigh these in relation to their habitual ethical frames, the gravity of their health realities, and the impact this will have on those who care for them. Subjective proximity is central to the decision-making process and the motivations behind seeking treatment abroad. The medical tourism industry has astutely tapped into what they know the patient seeks – affective care and medical treatment – a product that the tourist cannot procure at home and which prompts a search that is predicated by neoliberal health 'reform' and a feeling of consequent disenfranchisement at home.

A consequential model of ethical decision-making which focuses on the implications for those abroad (emergence of a two-tiered health system, erosion of public healthcare and systems, poorer healthcare for the poor) and the implications for their local health system (perpetuation of a two-tiered health system, erosion of public healthcare, poorer healthcare for the poor), for most of these people, would provoke ethical dissonance. Most would not be able to reconcile their habitual ethical and moral response – which is familial and individual (has subjective proximity), with an abstracted code of conduct that relates to an unrelated other who has no recognisable claim to their sense of obligation or responsibility. Furthermore, the 'ought' for these medical tourists (who are uninsured and cannot afford treatment at home) is not supported with an organisational 'can' locally, and there is no obvious articulator (or articulation) at an organisational level that would facilitate the development of new ethical dispositions. Indeed, their decision-making complies with the ethical norm – self-governance over healthcare, ability to procure healthcare, sound consumer choice (cheapest option (with reliability)), and putting the family first.

The problem with consequentialist models of ethical decision-making with respect to medical tourism for essential treatment, is that in failing to address wider societal inequalities which shape individual decision-making, resultant 'ethical pathways' request the disenfranchised not only to fend for themselves but also to put them 'selves' *after* other disenfranchised selves. It assumes the moral responsibility for preserving public health (which is essential for *all*, including the poor) rests with the unwell poor and those that are largely disenfranchised; and remains disturbingly silent on the role that might be played by the (substantially) wealthy and enfranchised minority.

Most research in the medical tourism field stresses the differences in wealth between medical tourists and locals in medical tourist destinations, but to date does not address the inequalities within those societies to which medical tourists seeking essential health services belong. Those seeking treatment abroad often cannot afford healthcare at home and their ethical and moral socialisation ensures that, if they can, they will seek care and treatment abroad. It remains the case that there are also those resident in the United States and Canada who *cannot* afford to seek treatment abroad. The curious thing about consequentialist models is that they obscure inequity within the so-called 'wealthy nations' and within simultaneously 'developing nations', stressing only the inequity *between* nations. Ensuring public healthcare is not eroded and remains available to all is a moral and ethical dilemma that all members of pluralistic societies are responsible for.

To provide ethical pathways for the unwell and those who cannot afford treatment at home, and not provide ethical highways for those who are well and can afford treatment, seems like a curious ethical solution to not only the inequities perpetuated by medical tourism but also to the erosion of public healthcare globally. While this is in part the problem of the impasse between particularism and universalism (Smith 2000: 15), the impasse cannot be addressed if the ethical nature of the particularism is not understood.

The key question underpinning this chapter was: what counts as ethical consumption in the field of medical tourism – and from whose perspective? From the perspective of these medical tourists seeking essential treatment in India their decision to travel for treatment conformed to their learned ethical competencies (frames). Subjectivity is central to how illness becomes linked to commodification and consumption. As Solomon (2012: 107) argues, affect is fundamental and constitutive of medical tourism (as it is to all healthcare), rather than ancillary. Our exploratory research, focusing on these postings, suggests that emotions, relationships, sentiment and recognition (affects) are entwined with habitual ethical frames that shape the decision-making processes of medical tourists who ultimately engage in a journey that offers hope and 'care'. Stewart speaks of ordinary affects pulling people into places where they didn't 'intend' to go (Stewart 2007) [recall the blogger who in response to unaffordable treatment at home said: 'I am not going to India']. Here, ordinary affects mediate and connect illness to consumer choice and consumption and habitual ethical frames (predominantly shaped by neoliberal philosophy) and offer these very experienced consumers the assurance that the decision is 'good' and 'right' and 'moral'.

Affective care is available in foreign hospitals; all that is needed is the ability to pay for it. The link between geographic space and healthcare, where proximity is key, has been eroded; not just by the speed of travel and the internet, but also by the neglect of public health provision in advanced capitalist societies such as the United States of America and Canada [and arguably elsewhere, in modified forms]. Medical tourism challenges the historic link between geographic proximity and healthcare provision and access, and ensures that *subjective proximity* becomes primary. The ability to circumvent affective neglect at home and seek affective care where it can be found – irrespective of physical distance – becomes the ethical [not new] and the emergent [and dominant] norm for those who have been disadvantaged by the neoliberal assault on public healthcare at home.

This is not to say that if medical tourism disadvantages others, this is right or good – but merely to highlight that there are a range of ethical frameworks that humans draw on, and in this instance choosing to go abroad for treatment complies with a frame that tells them it is 'right', 'good', and 'the best thing to do'. Sadly, everything that makes it 'right' and the 'best thing to do' is everything that has contributed to the erosion of public healthcare at home, their local disenfranchisement, and potentially the disenfranchisement of others. Empowerment is sought and gained abroad, rather than within the medical tourist's own borders.

Activity

1 Google search 'experiences of medical tourism'. Whose experiences are being recorded online?
2 Apply Snyder *et al.*'s ethical decision-making model (available in source noted below). Reflect on the analysis provided in this chapter and note down why this decision-making model has limitations when considering medical tourism as a consumptive practice.

Snyder, J., Crooks, V., Johnson, R. and Kingsbury, P. (2012) 'Beyond Sun, Sand, And Stitches: Assigning Responsibility For The Harms Of Medical Tourism' in *Bioethics*. Online. Blackwell publishing: Wiley OnLine Publishing. Available at: www.onlinelibrary.wiley.com

Recommended ethics reading

Fennell, D.A. (2006) *Tourism Ethics*, Clevedon: Channel View Publications.

For an overview of ethical issues in medical tourism, see:

Lovelock, B. and K.M. Lovelock (2013) *The Ethics of Tourism: Critical and Applied Perspectives*, London: Routledge.

Note

1 It is beyond the scope of this chapter to introduce a range of ethical theories which have applicability to consumption; for this reason the reader is referred to a number of useful sources at the end of the chapter.

References

Balaban, V. and Marano, C. (2010) 'Medical Tourism Research: A Systematic Review' in *International Journal of Infectious Diseases*, 14: e135 (Supplement 1) (14th International Congress on Infectious Diseases (ICID) Abstracts.

Banyai, M. and Glover, T.D. (2011) 'Evaluating Research Methods on Travel Blogs' in *Journal of Travel Research*, 51(3): 267–77.

Barnett, C., Cloke, Paul., Clarke, Nick and Malpass, Alice. (2005) 'Consuming Ethics: Articulating the Subjects and Spaces of Ethical Consumption' in *Antipode*, 37 (1): 23–45.

Bose, A. (2005) 'Private Health Care Sector In India: Is Private Health Care At The Cost Of Public Health Care? in *British Medical Journal*, 331:13: 38–9.

Burkett, L. (2007) 'Medical Tourism. Concerns, Benefits, And The American Legal Perspective' in *Journal of Legal Medicine*, 28: 223–45.

Connell, J. (2006) 'Medical Tourism: Sea, Sun, Sand And ... Surgery' in *Tourism Management*, 27: 1093–100.

Connell, J. (2011) *Medical Tourism*. Wallingford, Oxfordshire: CAB International.

Connell, J. (2012) 'Contemporary Medical Tourism: Conceptualisation, Culture And Commodification' in *Tourism Management*, 34: 1–13.

Cortez, N. (2008) 'Patient Without Borders: The Emerging Global Market For Patients And The Evolution Of Modern Healthcare' in *Indiana Law Journal*, 83: 71.

Crooks, V., Kingsbury, P., Snyder, J. and Johnston, R. (2010) 'What Is Known About The Patient's Experience Of Medical Tourism? A Scoping Review' in *BMC Health Services Research*, 10: 266.

Geddes, C.C., Henderson, A., MacKenzie, P. and Rodger, S.C. (2008) 'Outcome Of Patients From The West Of Scotland Travelling To Pakistan For Living Donor Kidney Transplants' in *Transplantation*, 86 (8): 1143–5.

Hall, C.M. (2011) 'Health And Medical Tourism: A Kill Or Cure For Global Public Health?' in *Tourism Review*, 66 (1–2): 4–15.

Hermans, H. (2000) 'Cross Border Health Care In The European Union, Recent Legal Implications Of Decker And Kohll' in *Journal of Evaluation in Clinical Practice*, 6: 431–9.

Jeevan, R. and Armstrong, A. (2008) 'Cosmetic Tourism And The Burden On The NHS' in *J Plast Reconstr Aesthet Surg*, 61 (12): 1423–4.

Johnston, R., Crooks, V.A., Snyder, J. and Kingsbury, P. (2010) 'What Is Known About The Effects Of Medical Tourism In Destination And Departure Countries? A Scoping Review' in *BMC Health Services Research*, 10: 266.

Johnston, R., Crooks, V.A. and Snyder, J. (2012) '"I Didn't Even Know What I Was Looking For": A Qualitative Study Of The Decision-Making Processes Of Canadian Medical Tourists' in *Globalization and Health*, 8: 23.

Krippendorff, K. (2004) *Content Analysis: An introduction to Its Methdology*, Thousand Oaks, CA: Sage Publications.

Lovelock, B. and Lovelock, K.M. (2013) *Ethics of Tourism: Critical and Applied Perspectives*, London: Routledge.

Lunt, N., Hardey, M. and Mannion, R. (2010) 'Nip, Tuck And Click: Medical Tourism And The Emergence Of Web-Based Health Information' in *The Open Medical Informatics Journal*, 4: 1–11.

Lunt, N. and Percivil C. (2011) 'Systematic Review Of Web Sites For Prospective Medical Tourists' in *Tourism Review*, 66 (1–2): 57–67.

Meghani, Z. (2011) 'A Robust, Particularist Ethical Assessment Of Medical Tourism' in *Bioethics*, 11 (1): 16–29.

Pearce, P.L. (2012) '"Tourists" Written Reactions To Poverty In Southern Africa' in *Journal of Travel Research*, 51: 154–65.

Pennings, G. (2007) 'Ethics without Boundaries: Medical Tourism' in R.E. Ashcroft, A. Dawson, H. Draper and J.R. McMillan (eds), *Principles of Health Care Ethics*, (2nd edition), Chichester: John Wiley and Sons Ltd.

Pocock, S. and Phua Kai Hong (2011) 'Medical Tourism And Policy Implications For Health Systems: A Conceptual Framework From A Comparative Study Of Thailand, Singapore And Malaysia' in *Globalization and Health*, 7: 12.

Smith, D.M. (2000) *Moral Geographies: Ethics in a World of Difference*, Edinburgh: Edinburgh University Press.

Smith, S. (2010) *Practical Tourism Research*, Wallingford, Oxfordshire: CAB International.

Smuddle, P.M. (2005) 'Blogging, Ethics And Public Relations: A Proactive And Dialogic Approach' in *Public Relations Quarterly*, 50 (3): 34–8.

Snyder, J., Crooks, V., Johnston, R. and Kinsbury, P. (2012) 'Beyond Sun, Sand, And Stitches: Assigning Responsibility For The Harms Of Medical Tourism' in *Bioethics*, 27 (5): 233–42.

Solomon, H. (2011) 'Affective Journeys: The Emotional Structuring Of Medical Tourism In India' in *Anthropology & Medicine*, 18 (1): 105–18.

Stewart, K. (2007). *Ordinary Affects*, Durham: Duke University Press.

Terzi, E., Kern, T. and Kohnen, T. (2008) 'Complications After Refractive Surgery Abroad' in *Ophthalmologe*, 105(5): 474.

Vuckovic, N. (1999) 'Fast Relief: Buying Time With Medications' in *Medical Anthropology Quarterly*, 13 (1): 51–68.

14 Marketing responsible tourism

Clare Weeden

Introduction

This chapter provides an overview of some of the key challenges that face those tasked with marketing responsible tourism. It begins with a brief review of the ethical issues that dominate discussion of tourism's impact, and an articulation of the main drivers for responsible tourism. The chapter defines responsible tourism, discusses some of the major barriers to increasing demand for responsible holidays, and reflects on the various marketing methods adopted by tour operators in this industry. The chapter concludes with recommendations for future consideration.

The success of mass tourism

Each year millions of people travel, take holidays and pursue leisure experiences in pursuit of self-discovery, relaxation and pleasure (Weeden 2013). Because such aspirations are often considered essential to life in post-industrial society, demand for travel has grown exponentially during the past 60 years. As a consequence, 4.8 billion international and domestic trips were taken in 2011 (UNWTO 2012a), while the forecasted increase in demand from the BRIC countries of Brazil, Russia, India and China means 1.8 billion international and seven billion domestic trips will be taken by 2030 (UNWTO 2011; UNWTO 2012b). Although tour operators may be delighted with such success and happy to meet the commercial opportunities presented by these forecasts, not everyone is pleased with the style of tourism's development, and not all stakeholders believe they benefit equally from tourist expenditure. For instance, while tourism undoubtedly brings economic benefit, some question whether an industry renowned for its propensity to offer high volume, low price tourism to the mass market can ever be truly sustainable and conform to socially just and equitable principles (Cleverdon and Kalisch 2000; Wearing 2002; Higgins-Desbiolles 2008).

In response to such criticism, commentators have spent decades debating various types of 'alternative tourism', discussing what these approaches might look like, and trying to establish how they could be sold successfully to the marketplace (see for example Eagles *et al.* 2002; Kalisch 2001; Scheyvens 2002). The most well-established of these proposed alternatives are ecotourism, community-based tourism, Pro-poor tourism, Fair Trade tourism, ethical tourism and responsible tourism. While debate continues as to the individual merits of these related concepts, tensions over terminology have detracted from the message that consumers and industry must move together towards the goal of taking responsibility in tourism (Cooper and Ozdil 1992). While clearly prioritising different aspects of tourism and travel, individually and collectively the aim of these alternatives is to foster a just and equitable approach to the development, operation and management of tourism. Such aspirations are critical to the future success of the industry, which is coming under 'increasing international pressure ... to address issues of global warming, social inequality and diminishing natural resources' (Frey and George 2008: 107). Indeed, calls have long been made for the tourism and travel industry to refocus attention on sustaining fair and cooperative relationships rather than relying on unjust appropriation of human and non-human capital (see for example Font and Ahjem 1998; Forsyth 1997; Hultsman 1995; Payne and Dimanche 1996). Responsibility in tourism and responsible forms of tourism are increasingly considered fundamental to the industry's future success (see Leslie 2012).

Defining responsible tourism

With its roots in the concept of sustainable tourism, responsible tourism has developed from the understanding that 'tourism-related actors can develop a sense of ethical and moral responsibility that has resonance beyond self-interest' (Bramwell *et al.* 2008: 253). This idea that tourism should deliver benefits to a range of stakeholders, and not just operate for the pleasure of the tourist, is inherent in the following definition,

> Responsible tourism minimises negative economic, environmental, and social impacts, generates greater economic benefits for local people and enhances the well-being of host communities. It aims to improve working conditions and access to the industry, involve local people in decisions that affect their lives and life chances, and make positive contributions to the conservation of natural and cultural heritage and to the maintenance of the world's diversity. Responsible tourism also strives to provide more enjoyable experiences for tourists through more meaningful connections with local people, and a greater understanding of local cultural, social and environmental issues. It provides access for physically challenged people, is culturally sensitive, and engenders respect between tourists and hosts so building local pride and confidence
>
> (International Centre for Responsible Tourism, n.d.)

Such priorities also underpin Goodwin and Pender's (2005) definition of responsible tourism, which although vague, echoes Cooper and Ozdil's (1992) point that shared responsibility is essential. In this instance they clarify the role of two key stakeholders – business and consumers. They assert responsible tourism is,

> [a] business and consumer response to some of the major economic, social and environmental issues which affect our world. It is about travelling in a better way and about taking responsibility for the impacts that our actions have socially and economically on others and on their social, cultural and natural environment.
>
> (Goodwin and Pender 2005: 303)

Much of the discussion about responsibility in business and consumer lifestyles can be traced back to George Fisk, who was one of the first to express concern over the global implications of mass production and consumption among the advanced nations of the world. His influential paper stressed individuals needed to take responsibility for limiting their personal consumption in order to encourage a rational and efficient (as opposed to extravagant and unnecessary) use of the world's resources (Fisk 1973). Like Goodwin and Pender, and also Cooper and Ozdil, Fisk called for all stakeholders (business, government and consumers) to work together on joint initiatives to curb excessive consumption. He cited Margaret Mead's understanding of responsibility in support of his point: '[R]esponsibility will include planning for lifestyles which are feasible economically and which will contribute to the sense of justice and dignity of all the people of the earth' (Mead 1970, as cited in Fisk 1973: 25). Mead's explicit mention of the connections between responsible consumption, social justice and human dignity, emerged much later – in the definitions of responsible tourism presented earlier, and in the principles of the Cape Town Declaration (2002) on responsible tourism in destinations, the latter of which underpins the ICRT's definition of responsible tourism.

Expectations that people and also corporations should take responsibility for their individual and collective acts are much more widespread now than they were in 1973, largely due to the United Nation's emphasis on sustainable development goals to be embedded in targets for global growth, and public demand for transparency in business. Such developments have been fundamental to the increasing adoption of corporate social responsibility (CSR) as a benchmark of ethical business practice. However, while commercial entities, as well as non-governmental organisations (NGOs), have largely accepted the need to embrace the concept of responsibility in tourism, it appears that positive consumer response to this is somewhat inconsistent.

For example, apart from Tourism Concern, a UK campaign group for ethical and fairly traded tourism, there is little evidence of consumer-led calls for tourists to adopt higher ethical standards when travelling. Indeed, given the small number of responsible holidays sold, and studies revealing people want to relax and have fun rather than think about ethical issues on holiday, it appears tourists are unaware of or uninterested in the negative impact of their holidays on other people, cultures and environments. While research indicates consumers are aware of the issues (see Mintel 2012), there is clearly a gap between tourists'

Consumer demand for responsible tourism

Given that tourism and travel industry actors are increasingly urged to take greater responsibility for their actions, and to rely less on others to deliver positive behavioural change (Coles *et al.* 2013), it is unfortunate that consumers are not rising to the challenge and playing their part to achieve this. As noted in the previous section, a significant challenge facing those keen to increase demand for responsible tourism is that while there is evidence of interest among both suppliers and tourists, the numbers of holidays sold remains tiny. For example, in 2011, UK-based responsible tour operators sold holidays worth £188 million. While this is an increase of 154 per cent since 1999 (and an increase of 54 per cent since 2008) (Cooperative Bank 2010, 2012), it is insignificant in comparison with the £35 billion UK outbound and domestic holiday market (Mintel 2012). With regard to consumers, more than 45 per cent of UK tourists do not want to think about ethical issues on holiday, and 49 per cent claim not to feel responsible for the environmental impact of their holidays (Mintel 2012).

Several reasons have been put forward as to why responsible tourism appears to be such a low priority for holidaymakers. These include consumers being confused about the array of labels (ethical, eco, sustainable, Fair Trade) used to describe responsible holidays. Also, while many tourists are adept at booking holidays independently, others may be challenged by the many inter-dependent components of the holiday business, and the sometimes-overwhelming amount of information needed to make what are often very complex bookings. Being required to read additional information about the ethical issues involved in their holidays increases the levels of decision-making for consumers, and may lead to information overload and subsequent disenfranchisement with the issue (Cherrier 2007). The desire to 'switch off' from the concerns of everyday life is documented as another reason as to why tourists disregard ethical issues in tourism (Barr *et al.* 2010). Similarly, as noted in ethical consumer studies (Szmigin *et al.* 2009), tourists are likely to prioritise convenience, quality and price over ethical considerations in their decision-making. Such issues further challenge those operators who are keen to deliver responsible holidays, a point already noted by the mass market as illustrated by the following quote,

> [Consumers will only] … purchase more sustainably if it is easy to do so, affordable and well communicated. [TUI] are very aware that destination quality, attractiveness, weather and comfort levels are key determinants of customer decision-making, the quality of holiday experiences and thus our business success.
>
> (Jane Ashton, TUI Travel's Director of Group Sustainable Development, 2010b: 11)

Some people ignore the ethical issues completely yet buy responsible holidays because they want to experience a quality holiday, travel in a small group and interact meaningfully with local communities, all of which responsible operators generally provide (Weeden 2002). The ethical component for such individuals is an incidental bonus rather than a deliberate element in their decision-making. In this instance, promotional campaigns emphasising quality and authenticity will be more relevant to consumers than marketing material that tries to reinforce the ethical component of the holiday. This may explain why some of the SMME operators mute their ethical credentials (Weeden 2005).

Indeed, companies who bring ethical brands to the market with no penalty in terms of quality or price are considered most likely to succeed as such criteria are at the top of most consumers' priorities (Page and Fearn 2005). Potentially, such an approach might widen appeal for responsible travel beyond the niche of the committed ethical consumer, as well facilitate the incorporation of sustainable values in tourism development. Such a recommendation would certainly correspond with the perspective that responsible tourism is a guiding philosophy and not a niche product. Unfortunately, moving the operations industry towards the situation whereby they offer responsibility as standard, rather than as an add-on or a competitive opportunity, is unlikely to happen in the short term.

Aside from the competitive barriers to encouraging the supply of responsible holidays, perhaps the most significant challenge in the drive to encourage responsible tourism stems from people's holiday motivations being extremely complex and highly subjective. They also reflect a range of physical, psychological and aspirational factors. From the small number of responsible holidays being sold, it appears few people are willing to supplement a potentially extensive list of personal motivations with an equally large number of ethical considerations. Also, many people who claim to be concerned about ethical issues in tourism (see Tearfund 2000a, 2000b, 2001) change their minds, or do not act upon their intentions at the point of purchase (Budeanu 2007; Choi and Sirakaya 2005; Hares *et al.* 2010; Miller *et al.* 2010; Sharpley 2000).

This gap between attitudes, intentions and behaviour has been extensively examined across all areas of the ethical consumer literature (Shaw and Clarke 1999; Shaw and Shiu 2003). It has also been addressed in tourism. For example, studies of consumer attitudes with regard to climate change and personal air travel reveal that holiday decision-making is perceived differently from that associated with everyday purchases. Using social practice theory, this research indicates holidays are considered 'sacred' and explained as hard habits to break (Barr *et al.* 2010; Dickinson *et al.* 2010; Verbeek and Mommaas 2008). Even consumers who are environmentally focused in everyday life rationalise their less-than-ethical holiday choices with statements like, 'it is too difficult for me to change my behaviour', 'my actions as an individual won't make any difference', and 'I protect the environment in other ways' (see Stoll-Kleemann *et al.* 2001: 112). Such studies are not the focus of this chapter. Indeed, detail of consumer attitudes and their impact on behaviour can be found in greater detail elsewhere (Carrington *et al.* 2010). Rather, the focus of this chapter is to discuss how tour operators can use such information to market and promote responsible tourism. The following sections examine this in greater depth.

Tour operators and responsible tourism

The previous section discussed some of the challenges facing operators who want to increase demand for responsible tourism. Indeed, through their supply of holiday and leisure experiences, tour operators are pivotal in the development of a responsible industry, and should be encouraged to do more to achieve this objective. Unfortunately, mass market tour operators are often considered part of the problem, and the 'bad boys' of tourism. However, while they have been pilloried for destroying the social, cultural and environmental resources on which they rely (see McKercher 1993), there is evidence that many tour operators are passionate about delivering a responsible product (see Weeden 2005).

Arguably, a more accurate assessment of operators' potential contribution towards this goal would acknowledge the diverse nature of the different operations, both in size and product offer, the financial imperatives of conducting business in such a highly competitive market, and the extensive efforts of many operators, not only to demonstrate their ethical credentials, but also to facilitate change in others. Admittedly, the majority of the latter tend to be small-, medium- and micro-sized enterprises (SMMEs), whose owner-managers claim to feel a personal moral obligation to care about the impacts of their products and services (see Weeden 2005). The most prominent of such organisations operating out of the UK include Tribes Travel, Discovery Initiatives, Rainbow Tours, Steppes Discovery, Baobab Travel and Exodus Travel. Such companies are not confined to the UK – similarly motivated companies operate in all areas of the world.

Having noted the strong emphasis on responsibility from SMMEs operating in tourism, who rely on access to quality resources to compete effectively in an oligopolistic marketplace, there is also some evidence that the mass market is adapting its activities in line with calls for a more just and equitable system. Admittedly the evidence is patchy, but Europe's largest operator, TUI Travel, arguably deserves credit for attempting to promote responsibility and sustainable practice. Widely recognised for their commitment to sustainable tourism, responsible leadership strategy, and carbon reduction initiatives, TUI focuses on promoting a positive business case for responsibility in order to encourage other operators to follow suit. The company claims that acting responsibly enables them to attract investors, reduce costs, recruit quality staff, and improve the quality of their products (TUI Travel 2010a). In combination, these benefits allow TUI to gain competitive advantage, which in turn enables them to spend money and time on protecting their access to social, cultural and environmental assets.

Cynics might argue such pronouncements are self-serving and merely marketing and public relations puffery, especially in light of TUI's First Choice brand controversially moving exclusively to all-inclusive hotels in summer 2012. However, it would be unfair to blame a single set of stakeholders for the relative status quo in the responsible tourism market, especially as commercial organisations, by definition, can only survive if they successfully fulfil customer needs. Indeed, tour operators who understand the wider importance of promoting responsible holidays face considerable challenge – not only must they secure profitable sales in an increasingly competitive global market, they also have to ensure supply chains conform to corporate promises on responsibility, and persuade consumers of the benefits of incorporating responsibility into their holiday decisions.

It is this latter issue that is one of the major obstacles to those seeking to develop responsible tourism beyond its somewhat niche status. Indeed, if European tourists are not interested in buying responsible holidays, the industry must find new ways to encourage people to change, not only their attitudes, but also their behaviour. Such an objective has been discussed extensively in the climate change and tourism debate, with social practice theory being mooted as one explanation for the attitude–behaviour gap, as noted earlier (see Barr *et al.* 2010; Hares *et al.* 2010). Much of the literature generated by these studies also identifies the potential of marketing to positively influence consumer behaviour change, but this remains relatively little-discussed in terms of tour operators. Such an emphasis is the key focus of the following section, which discusses how operators commonly use marketing to promote responsible holidays, and considers some of the challenges associated with such activity.

Using the principles of marketing to sell ethical products

With its roots in 1950s USA, marketing is a critical management function and an essential component in the effective promotion of products, services, ideologies and ideas (Belz and Peattie 2012). It has been defined as 'organised effort, activity and expenditure designed first to acquire a customer and second to maintain and grow a customer at a profit' (Kotler *et al.* 2009: 4). Similarly, the American Marketing Association clarifies it as the 'activity, set of institutions, and processes for creating, communicating, delivering, and exchanging offerings that have value for customers, clients, partners and society at large' (in Pomering *et al.* 2011: 958).

Unfortunately, given its pivotal role in influencing demand, society holds marketing and marketing professionals in very low esteem. Marketing activity is often perceived to be unethical, with marketers held responsible for consumers being persuaded to buy products and services they do not need, and for exploiting not only the people buying the products but also those producing them (Neyland and Simakova 2009). Those tasked with marketing products that have an ethical dimension (like responsible tourism) face even greater challenge because consumers also exhibit extreme cynicism about the inherently unethical nature of the corporate world. Such mistrust and scepticism is in large part due to increasing (internet) access to news media, citizen-consumer reports of dubious marketing practice, and the extensive levels of ethical and green wash in the 1980s (Peattie and Crane 2005).

How much people distrust an organisation that claims to be ethical seems to correspond with its size. For instance, mistrust is particularly prevalent when the vendor organisation is a large multinational corporation (de Pelsmacker et al. 2005). Arguably it is attitudes such as these that encourage many ethically motivated organisations to play down their ethical, environmental and/or CSR achievements (Crane 1997). Indeed, companies with proactive CSR policies are often reticent to proclaim them for fear of being accused of exploiting moral obligation for the (cynical) pursuit of profit (Steger et al. 2007; van de Ven 2008).

In light of such perceptions, and in tandem with the rise of the citizen-consumer, marketing professionals are increasingly acknowledging the need to refocus their efforts to develop a sustainable approach to marketing. Initiated by Fisk's concern about excessive consumption levels, and influenced – as noted earlier – by the emphasis on sustainable development in the United Nations's Brundtland Report (World Commission on Environment and Development (WECD) 1987), sustainable marketing has been developed with the intention of combining the following: satisfying customer needs, meeting organisational goals, being compatible with ecosystems, and cognisant of society's long-term needs (Fuller 1999). Sustainable marketing in tourism places emphasis on sustainability at each stage in the process of marketing, on the premise that marketing is 'a way of doing business rather than a mere managerial function' (Pomering et al. 2011: 955) These authors argue that traditional marketing mixes do not address broader societal concerns, and believe,

> The concept of the marketing mix is the ideal checklist for examining what tourism organisations currently do and how they might more appropriately meet increasing sustainability demands. The elements of the marketing mix are captured in the core values of the organisation, reflecting the nature of its relationships with key stakeholders, such as suppliers, consumers, employees, host communities and the environment. These relationships signal the degree of the organisations' sustainability orientation.
> (Pomering et al. 2011: 961)

In common with responsible tourism, sustainable marketing has spawned a number of labels including green, ecological, social and societal marketing, which to some extent has resulted in confusion and also misunderstanding (Belz and Peattie 2012; Fuller 1999). Also, like responsible tourism, it has been the focus of debate, with some arguing the term is an oxymoron (Fuller, 1999), and asserting 'marketing [is] traditionally considered an enemy of sustainability' (Pomering et al. 2011: 953). This latter point is critical to those promoting responsible holidays. Indeed, tourism marketing has been criticised for focusing too much on opportunities for economic growth and too little on the impact of tourism development on people, places and the environment (Jamrozy 2007). It is also blamed for encouraging tourists to believe greater levels of 'happiness' can be attained with each additional holiday purchase, a theme evident in many areas of modern life (Kasser 2004).

Social marketing in tourism

It is clear from the discussions in this chapter so far that marketing is often perceived negatively. As such, it is not surprising that those who use it for professional reasons have sought to ally themselves with a more sustainable remit. However, it is far from the only initiative tasked with the rehabilitation of marketing as a socially positive function. For example, although it shares similar objectives with sustainable marketing, social marketing has lately been lauded as the most effective way of influencing, changing and/or preventing behaviour (see Belz and Peattie 2012). It is most commonly observed in campaigns associated with health, where it is used to discourage harmful behaviour, such as drug and alcohol addiction, or to promote positive physical activity aimed at improving individual health and wellbeing.

In its purest form, the primary difference for social marketing is that it aims to benefit targeted individuals, groups and societies, whereas the objective of traditional marketing is to benefit the organisation carrying out the promotional activity (Andreasen 1994). Social marketing, a term first coined by Kotler and Zaltman in 1971, has been defined as, 'the adaptation of commercial marketing technologies to programmes designed to influence the voluntary behaviour of target audiences to improve their personal welfare and that of the society of which they are a part' (Andreasen 1994: 110). While using similar techniques, social marketing differs from traditional marketing because of an emphasis on,

> understanding and overcoming barriers to behavioural change. Where a commercial marketing approach might begin by asking why people might buy this coffee, and how they can be encouraged to do so, a social marketing approach would first seek to understand why most do not buy it, and how their objections might be overcome ... for a product seeking to move beyond a market niche, tackling the reason for non-purchase amongst the majority of consumers is crucial.
>
> (Golding and Peattie 2005: 160)

The utility of social marketing to influence consumer demand for responsible tourism has been noted for some time. For example, it is more than a decade since Dinan and Sargeant (2000) first suggested ecotourism destinations could adopt their techniques to target tourists already predisposed to act sustainably. Significantly, however, it is also critical if seeking to activate 'new' patterns of behaviour, making it useful for widening demand for responsible tourism by reaching out to tourists not yet convinced of the need to adopt such behaviour. Indeed, given the prevalence of the attitude–behaviour gap, whereby consumers express concern but take no action, a technique that encourages people to change their behaviour is definitely worth exploring, not only to increase consumer demand for responsible holidays but to persuade the industry to adopt more responsible forms of development.

This is the key challenge for any initiative aimed at modifying human behaviour; people are notoriously reluctant to abandon their habitual life patterns (Dinan and Sargeant 2000), particularly if addiction is involved. This is highly relevant in tourism and travel, where travellers have been labelled as displaying an addiction to flying (Cohen et al. 2011). Although this might initially be interpreted as an extreme view, behaviour commonly associated with addiction, such as patterns of denial, suppression of guilt, an off-loading of responsibility, and failed attempts to end destructive behaviour, certainly has resonance in tourist research and an improved understanding of addictive behaviours may contribute to greater knowledge of the factors involved in the attitude–behaviour gap. Certainly the conceptualisation of tourist inertia (express concern but do nothing) as an 'addiction' merits further research attention (for more debate on this issue see Gössling 2013).

Of course, no single solution will ever be sufficient to overcome the complex set of challenges faced by those wanting to promote responsible tourism to society, and social marketing is no exception to this. Indeed, while it may assist in communicating the benefits of tourism, and promoting social equity and international understanding through its operation (Jamrozy 2007), there are several reasons why it might be difficult to encourage the travel industry to fully endorse the techniques associated with it. First, social marketing is considered by many to be the remit of public and non-profit organisations rather than the private sector. Second, a key objective of social marketing is to enhance society's wellbeing. This would be challenging for any tour operator, especially a public company, as investors may be less interested in social benefits and more concerned with share dividends. Third, when the objects of the campaign are in a far-off country, as they are likely to be for many holidays, the object of the campaign is so far removed from the consumer that it will be a challenge to make the connection meaningful (Weeden 2013).

Conclusion

Given the extensive list of challenges so far discussed in this chapter regarding the marketing of ethical products and the related problems of incorporating ethical messages in holiday promotion, how might the industry move forward to embrace a sustainable future, at the same time as encourage a greater demand for responsible travel? There are of course many group initiatives already in existence, which have been formed specifically with these joint objectives in mind. For example, Tourism Concern's Ethical Tour Operator Group (ETOG) focuses on providing support and encouragement to tourism SMEs to provide ethical and fairly traded tourism. A different programme, the Tour Operators' Initiative (TOI), involves a range of tour operators, including TUI plc and Accor, who work together with UNEP and UNWTO to support sustainable tourism development.

With regard to the tourist, Stanford (2008: 270) suggests appealing to those who already exhibit elements of responsible consumption economically, environmentally or culturally, 'to demonstrate a greater degree of responsibility within the dimensions they already practice'. This group of consumers could be potentially significant in terms of increasing demand for responsible tourism. For example, while nearly half of UK consumers do not feel responsible for the impact of their holidays, 32 per cent are unsure, with a similar number claiming not to know whether they want to think about ethics on holiday (Mintel 2012). The existence of such a group, and their potential for persuasion, has also been recognised by TUI,

> [A]lthough there is not an active majority that consider sustainability as an integral part of their decision-making process, industry should look promisingly at the large number of ambivalent consumers that are within the field of influence. Consumer demographics demonstrate divergence in attitudes towards sustainability amongst different consumer groups but encouragingly, no particular group in terms of age or lifestyle.
> (Jane Ashton, Director for Sustainable Development, TUI Travel, 2010: 11)

Such ideas have their roots firmly in the marketing domain, where efforts to increase usage (and therefore purchase of a product or service) from light to heavy are seen as essential in campaigns to cost-effectively increase demand. However, and this is the nub of the issue – how might the tour operations industry collectively move tourists towards an expectation of responsibility from operators, while encouraging them to accept they also have a personal role to play? The industry is notoriously competitive and run on very tight financial margins. Such extreme trading conditions almost inevitably lead to power struggles between all operators, large and small. While much of the literature on responsible tourism has called for collaboration and cooperation among all stakeholders, the industry at present shows little sign of moving towards this – apart from voluntary initiatives, the competition is as tight as ever.

Consumers believe they benefit from such competition, not least because of the impact it has on the price they pay for their holidays. Competition does not, however, automatically increase quality, and quality holidays will be the domain of future developments in sustainability and responsible tourism. Unless everyone, supplier and consumer alike, works together to achieve such an objective, a truly responsible industry will be difficult to secure. Tourists remain at the heart of the tourism industry and so efforts to encourage them to embrace responsible tourism are essential. While all tour operators ought to be more proactive in the development of a responsible industry, every industry actor needs to do more to deliver leisure and business experiences that provide beneficial relationships and positive financial outcomes for all, not only to secure the future of the industry but also to protect the social, cultural and natural assets on which they depend.

Activity

Ask students to read the following:

Brunet, S., Bauer, J., De Lacy, T. and Tshering, K. (2001) 'Tourism Development in Bhutan: Tensions between tradition and modernity' in *Journal of Sustainable Tourism*, 9 (3): 243–63.

Use the article to discuss the questions below.

Also, ask students to collect, and bring to the class, images of Bhutan, either online or from tour operator brochures. They can use these to understand how Bhutan is marketed to international tourists, as well as familiarise themselves with its physical and cultural geography. Make sure they know where in the world the country is located!

1 What are the key objectives for Bhutan's tourism development?
2 How does Bhutan control tourism to and within the country?
3 What do you consider to be the likely tensions between international tourists' needs and expectations and Bhutan's desire to maintain its unique cultural identity?
4 How can Bhutan's policy of the 'Middle path' to gross national happiness be useful to other nations with similarly vulnerable cultures?
5 Use additional sources to determine whether Bhutan has modified its attitude to tourism development since the article was published in 2001

Additional resources

www.bbc.co.uk/news/world-south-asia-12480707 www.chinadialogue.net/article/show/single/en/5025-Can-Bhutan-forge-a-new-tourist-path-
www.sustainabledevelopment.un.org/content/documents/798bhutanreport.pdf

References

Andreasen, Alan R. (1994) 'Social Marketing: Definition and Domain' in *Journal of Marketing and Public Policy*, Spring: 108–14.
Barr, S., Shaw, G., Coles, T. and Prillwitz, J. (2010) 'A Holiday is a Holiday: Practicing Sustainability, Home and Away' in *Journal of Transport Geography*, 18 (3): 474–81.
Belz, F.-M. and Peattie, K. (2012) *Sustainability Marketing: A Global Perspective*, 2nd edition, Chichester: John Wiley & Sons Ltd.
Bramwell, B., Lane, B., McCabe, S., Mosedale, J. and Scarles, C. (2008) 'Research Perspectives on Sustainable Tourism' in *Journal of Sustainable Tourism*, 16 (3): 253–7.
Budeanu, A. (2007) 'Sustainable Tourist Behaviour: A Discussion of Opportunities for Change' in *International Journal of Consumer Studies*, 31 (5): 499–508.
Cape Town Declaration (2002) 'Responsible Tourism in Destinations'. Online, available at: www.responsibletourismpartnership.org/CapeTown.html (accessed 12 June 2013).
Carrington, M.J., Neville, B.A. and Whitwell, G.J. (2010) 'Why Ethical Consumers Don't Walk their Talk: Towards a Framework for Understanding the Gap Between the Ethical Purchase Intentions and Actual Buying Behaviour of Ethically Minded Consumers' in *Journal of Business Ethics*, 97 (1): 139–58.
Cherrier, H. (2007) 'Ethical Consumption Practices: Co-production of Self-Expression and Social Recognition' in *Journal of Consumer Behaviour*, 6 (5): 321–35.
Choi, H.S. and Sirakaya, E. (2005) 'Measuring Resident Attitudes Toward Sustainable Tourism: Development of a Sustainable Tourism Attitude Scale' in *Journal of Travel Research*, 43(4): 380–94.

Cleverdon, R. and Kalisch, A. (2000) 'Fair Trade in Tourism' in *International Journal of Tourism* Research, 2 (3): 171–87.

Cohen, S.A., Higham, J.E.S. and Cavaliere, C.T. (2011) 'Binge Flying: Behavioural Addiction and Climate Change' in *Annals of Tourism Research*, 38 (3): 1070–89.

Coles, T., Fenclova, E. and Dinan, C. (2013) 'Tourism and Corporate Social Responsibility: A Critical Review and Research Agenda' in *Tourism Management Perspectives*, 6: 122–41.

Cooper, C.P. and Ozdil, I. (1992) 'From Mass to "Responsible" Tourism: The Turkish Experience' in *Tourism Management*, 13 (4): 377–86.

Cooperative Bank (2010) 'Ten Years of Ethical Consumerism: 1999–2008'. Online, available at: www.goodwithmoney.co.uk (accessed 11 June 2010).

Cooperative Bank (2012) 'Ethical Consumers Market Report 2012'. Online, available at: www.co-operative.coop/corporate/Investors/Publications/Ethical-Consumerism-Report/ (accessed 4 January 2013).

Crane, A. (1997) 'The Dynamics of Marketing Ethical Products: A Cultural Perspective' in *Journal of Marketing Management*, 16 (6). 561–77.

De Pelsmacker, P., Janssens, W., Sterckx, E. and Mielants, C. (2005) 'Consumer Preferences for the Marketing of Ethically Labelled Coffee' in *International Marketing Review*, 22 (5): 512–30.

Dickinson, J.E., Robbins, D. and Lumsdon, L. (2010) 'Holiday Travel Discourses and Climate Change' in *Journal of Transport Geography*, 18 (3): 482–9.

Dinan, C. and Sargeant, A. (2000) 'Social Marketing and Sustainable Tourism: Is There a Match?' in *International Journal of Tourism* Research, 2 (1): 1–14.

Eagles, P.F.J., McCool, S.F. and Haynes, C.D.A. (2002) 'Sustainable Tourism in Protected Areas: Guidelines for Planning and Management', Switzerland and Cambridge: IUCN Gland. Online, available at: www.cmsdata.iucn.org/downloads/pag_008.pdf (accessed 16 June 2013).

Fisk, G. (1973) 'Criteria for a Theory of Responsible Consumption' in *Journal of Marketing*, 37 (2): 24–31.

Font, X. and Ahjem, T.E. (1999) *Searching for a Balance in Tourism Development Strategies*, MCB Virtual Conference Centre, MCB University Press.

Forsyth, T. (1997) 'Environmental Responsibility and Business Regulation: The Case of Sustainable Tourism' in *The Geographical Journal*, 16 (3): 270–80.

Frey, N. and George, R. (2008) 'Responsible Tourism and the Tourism Industry: A Demand and Supply Perspective' in A. Spenceley (ed.), *Responsible Tourism: Critical Issues for Conservation and Development*, Gateshead: Earthscan: 107–28.

Fuller, D.A. (1999) *Sustainable Marketing: Managerial-Ecological Issues*, Thousand Oaks, CA: Sage Publications Inc.

Golding, K. and Peattie, K. (2005) 'In Search of a Golden Blend: Perspectives on the Marketing of Fair Trade Coffee' in *Sustainable Development*, 13 (3): 154–65.

Goodwin, H. and Pender, L. (2005) 'Ethics in Tourism Management' in L. Pender and R. Sharpley (eds), *The Management of Tourism*, London: Sage Publications: 288–304.

Gössling, S. (2013) 'Advancing a Clinical Transport Psychology' in *Transportation Research Part F: Traffic Psychology and Behaviour*, 19: 11–21.

Hares, A., Dickinson, J. and Wilkes, K. (2010) 'Climate Change and the Air Travel Decisions of UK Tourists' in *Journal of Transport Geography*, 18 (3): 466–73.

Higgins-Desbiolles, F. (2008) 'Justice Tourism and Alternative Globalisation' in *Journal of Sustainable Tourism*, 16 (3): 345–64.

Hultsman, J. (1995) 'Just Tourism, an Ethical Framework' in *Tourism Management*, 11 (9): 553–67.

ICRT (n.d.) 'Responsible tourism'. Online, available at: www.icrtourism.org/responsible-tourism/ (accessed 27 June 2013).

Jamrozy, U. (2007) 'Marketing of Tourism: A Paradigm Shift Toward Sustainability' in *International Journal of Culture, Tourism and Hospitality Research*, 1 (2): 117–30.

Kalisch, A. (2001) *Tourism as Fair Trade, NGO Perspectives*, London: Tourism Concern.

Kasser, T. (2004) 'The Good Life or the Goods Life? Positive Psychology and Personal Well-being in the Culture of Consumption' in P.A. Linley and S. Joseph (eds), *Positive Psychology in Practice*, Hoboken, New Jersey: John Wiley and Sons Inc.: 55–67.

Kotler, P., Keller, K.L., Brady, M., Goodman, M. and Hansen, T. (2009) *Marketing Management*, Harlow: Pearson Education Ltd.

Leslie, D. (ed.) (2012) *Responsible Tourism: Concepts, Theory and Practice*, Wallingford, Oxfordshire: CAB International.

McKercher, B. (1993) 'Some Fundamental Truths About Tourism: Understanding Tourism's Social and Environmental Impacts' in *Journal of Sustainable Tourism*, 1 (1): 6–16.

Miller, S. and Gregan-Paxton, J. (2006) 'Community and Connectivity: Examining the Motives Underlying the Adoption of a Lifestyle of Voluntary Simplicity' in *Advances in Consumer Research*, 33: 289.

Mintel (2012) Holiday Review – UK – January 2013.

Neyland, D. and Simakova, E. (2009) 'How Far Can we Push Sceptical Reflexivity? An Analysis of Marketing Ethics and the Certification of Poverty' in *Journal of Marketing Management*, 25 (7–8): 777–94.

Nunkoo, R. and Ramkissoon, H. (2010) 'Gendered Theory of Planned Behaviour and Residents' Support for Tourism' in *Current Issues in Tourism*, 13 (6): 525–40.

Page, G. and Fearn, H. (2005) 'Corporate Reputation: What do Consumers Really Care About?' in *Journal of Advertising Research*, 45 (3): 305–13.

Payne, D. and Dimanche, F. (1996) 'Towards a Code of Conduct for the Tourism Industry: An Ethics Model' in *Journal of Business Ethics*, 15 (9): 997–1007.

Peattie, K. and Crane, A. (2005) 'Green Marketing: Legend, Myth, Farce or Prophesy?' in *Qualitative Market Research: An International Journal*, 8 (4): 357–70.

Pomering, A., Nobel, G. and Johnson, L.W. (2011) 'Conceptualising a Contemporary Marketing Mix for Sustainable Tourism' in *Journal of Sustainable Tourism*, 19(8): 953–69.

Scheyvens, R. (2002) *Tourism for Development: Empowering Communities*, Harlow: Prentice Hall.

Sharpley, R.A.J. (2000) 'Tourism and Sustainable Development: Exploring the Theoretical Divide' in *Journal of Sustainable* Tourism, 8 (1): 1–19.

Shaw, D. and Clarke, I. (1999) 'Belief Formation in Ethical Consumer Groups: An Exploratory Study' in *Marketing Intelligence and Planning*, 17 (2): 109–19.

Shaw, D. and Shiu, E. (2003) 'Ethics in Consumer Choice: A Multivariate Modelling Approach' in *European Journal of Marketing*, 37 (10): 1485–98.

Stanford, D. (2008) ' "Exceptional" Visitors: Dimensions of Tourist Responsibility in the Context of New Zealand' in *Journal of Sustainable Tourism*, 16 (3): 258–75.

Steger, U., Ionescu-Somers, A. and Salzmann, O. (2007) 'The Economic Foundations of Corporate Sustainability' in *Corporate Governance*, 7 (2): 162–77.

Stoll-Kleemann, S., O'Riordan, T. and Jaeger, C.C. (2001) 'The Psychology of Denial Concerning Climate Mitigation Measures: Evidence from Swiss Focus Groups' in *Global Environmental Change*, 11: 107–17.

Szmigin, I., Carrigan, M. and McEachern, M.G. (2009) 'The Conscious Consumer: Taking a Flexible Approach to Ethical Behaviour' in *International Journal of Consumer Studies*, 33 (2): 224–31.

Tearfund (2000a) 'A Tearfund Guide to Tourism: Don't Forget your Ethics!' Online, available at: www.tearfund.org (accessed 11 January 2000).

Tearfund (2000b) 'Tourism: An Ethical Issue', January. Online, available at: www.tearfund.org (accessed 11 January 2000).

Tearfund (2001) 'Tourism: Putting Ethics into Practice', January. Online, available at: www.tearfund.org (accessed 20 February 2001).

TUI Travel (2010a) 'The Business Case for Sustainable Tourism'. Online, available at: www.tuitravelplc.com/sustainability/in-focus/business-case-sustainable-tourism (accessed 13 June 2013).

TUI Travel (2010b) 'Tourism Future'. Online, available at: www.tuitravelplc.com (accessed 13 June 2013).

UNWTO (2011) 'Tourism Towards 2030: Global Overview'. Online, available at: www.pub.unwto.org//111014_TT_2030_global_overview_excerpt.pdf (accessed 20 November 2012).

UNWTO (2012a) 'International Tourism to Reach one Billion in 2012'. Online, available at: www.media.unwto.org/en/press-release/2012-01-16/international-tourism-reach-one-billion-2012 (accessed 3 October 2012).

UNWTO (2012b) 'Presentation: World Tourism Performance 2011 and Outlook 2012'. Online, available at: www.media.unwto.org/en/press-release/2012–01–16/international-tourism-reach-one-billion-2012 (accessed 20 November 2012).

Van de Ven, B. (2008) 'An Ethical Framework for the Marketing of Corporate Social Responsibility' in *Journal of Business Ethics*, 82 (2): 339–52.

Verbeek, D. and Mommaas, H. (2008) 'Transitions to Sustainable Tourism Mobility: The Social Practices Approach' in *Journal of Sustainable Tourism*, 16 (6): 629–44.

Wearing, S. (2002) 'Re-centring the Self in Volunteer Tourism' in G.S. Dann (ed.), *The Tourist as a Metaphor of the Social World*, Wallingford, Oxfordshire: CAB International, 237–62.

WECD (1987) *The Brundtland Report (Our Common Future)*, Oxford: Oxford University Press.

Weeden, C. (2002) 'Ethical Tourism: An Opportunity for Competitive Advantage?' in *Journal of Vacation Marketing*, 8 (2): 141–53.

Weeden, C. (2005) 'Ethical Tourism: Is its future in Niche Tourism?' in M. Novelli (ed.), *Niche Tourism*, Oxford: Butterworth-Heinemann: 233–45.

Weeden, C. (2013) *Responsible Tourist Behaviour*, London: Routledge.

15 Concluding remarks

Karla Boluk and Clare Weeden

The following chapter discusses three discernible areas. Initially it summarises the current scholarship in the area of tourism and ethics. Second, the chapter reviews and highlights some of the significant contributions made in the preceding chapters, focusing explicitly on the compromises and tensions faced by travellers making ethical decisions in tourism. Last, the chapter reviews opportunities for future research on the topic of ethical consumption in tourism.

As stated in the introduction, ethical discourse has become widespread, and so it is no surprise to see that it has entered the field of tourism. Over a decade ago, Stark (2002) put forth that the subject of ethics in relation to the economy, the environment and society were mainly discussed in the concluding remarks of research papers encouraging areas for future potential research. This has also been the case in the discipline of tourism, until recently. There are a variety of reasons why ethics in tourism has been largely overlooked. For many, ethics is abstract, subjective and perceived as challenging to engage in, and therefore daunting. When ethical debate emerged in the discipline of tourism, discussions on the environment seemed to overshadow social issues, perhaps prompted by the Brundtland Report of 1987. Some researchers, such as Fennell (2008: 223), have highlighted the significance of focusing on ethics in tourism, arguing an in-depth exploration of ethics is imperative for the 'future advancement of tourism'. In this way, discussions have potential to develop a new pedagogy within the industry, confronting and challenging the negative developments that have impacted international stakeholders. The ethical dialogue that has begun to emerge in the discipline of tourism has largely developed from an understanding of the impacts of tourism development (see Smith and Duffy 2003). However, there is also support among many researchers for alternative approaches that spread the benefits and create opportunities for local stakeholders involved in and/or affected by tourism.

Early research into ethics in tourism, although not explicit, tended to focus on tourism conduct. For instance, Krippendorf (1984, 1991) argued a re-education in tourism was necessary to redress the negative impacts of tourism, which led directly to campaigns encouraging tourists to be mature, think critically, and assume responsibility when travelling (see for example, promotions by Voluntary Service Overseas (VSO) and Tearfund (in Weeden 2002)). Moving forward,

Poon (1993, 2003) conceptualised 'new tourism', which in turn opened the door to a plethora of alternative tourism approaches concerned with sustainability and responsible conduct. Accordingly, new tourism catered for an 'enlightened traveller', one who was interested in interacting with the cultures and peoples encountered. Both Krippendorf and Poon emphasised the importance of responsible conduct for both tourists and operators, and so it is clear that ethical discussions in tourism are not new. More recently, Pritchard *et al.* (2011) articulate some of the contradictions in tourism, which place heightened pressure on travellers to compromise between a much-deserved break from contemporary lifestyle demands and their intentions to 'be good'. For example, they argue that 'tourism worlds are worlds of ugliness-beauty, pain-pleasure, toil-relaxation, poverty-luxury, fear-comfort, hate-love, sacredness-profanity, and despair-hope' (Pritchard *et al.* 2011: 957). Such binary positions inevitably influence both the process and progress of ethical decision-making in tourism.

Our goal for this edited volume was to move the debate forward by examining some of the barriers and conspicuous tensions that stand in the way of such progress. Specifically, we wanted to give space to scholars who were actively engaged with research about ethical practice in tourism, and whose studies would provide support for conscientious travellers and tourism providers who were keen to engage more fully with ethical consumption in tourism. We also wanted the book to connect with and gently 'nudge' those who may be interested in making informed decisions but do not yet have the knowledge or time to research ethical alternatives. As such, the preceding 15 chapters have provided a uniquely comprehensive overview of the various challenges faced by contemporary travellers in the process of ethical consumption. The fundamental aim of this volume was to shed light on some of the compromises and tensions experienced by travellers, which created barriers to the pursuit of ethical consumption. Furthermore, the book intended to initiate dialogue on the challenges and opportunities for ethical consumption in tourism moving forward. As such, the 17 researchers who have contributed to, and made this volume possible, have demonstrated their expertise in the subject area by examining and teasing out many of the complexities and wide ranging issues that ethical consumption in tourism creates in both established and emerging areas.

The book has been organised into three main parts and designed to reflect the key debates. Specifically, these include the challenges of consuming ethically in tourism, and the complexities of providing support for travellers seeking to make more informed decisions while planning and carrying out their holidays. Analysis in the area of ethical consumption in tourism is emerging, yet, not surprisingly, what the chapters demonstrate are the variety of issues that are touched upon and often reiterated throughout the different chapters; the most common elements of which recognise the pervasive tensions and compromises that travellers must negotiate within the field of ethical consumption in tourism. Such concerns not unexpectedly reflect the organisation of the book and span all contributions.

For instance, several chapters reflect the fundamental difficulty associated with coupling ethics with tourism: the process of consumption in tourism necessarily invokes tension between satisfying and maximising one's desires and the imperative to do 'what is right'. Such challenges are apparent in all discussions regarding tourism consumption and the promotion of sustainability and ethical conduct on holiday. A growing awareness among consumers regarding the consequences of their consumption is addressed in several of the chapters. However this was at odds with a desire to relax and have fun. Specifically, Andrew Holden (Chapter 5) establishes that environmental and/or a conservationist ethic can stand in the way of and/or motivate travellers to make informed choices concerning the environment.

Not surprisingly, travellers pick and choose their compromises and behavioural change. Adriana Budeanu and Tareq Emtairah (Chapter 6) establish that the condition needed to support a more sustainable future is to present an *Environmentally Preferred Option* to travellers in light of their transportation, accommodation and leisure needs. Similarly, Michael Hall (Chapter 3) argues incremental change is possible within the tourism industry, and the provision of options and the utilisation of libertarian paternalism and gentle nudging as described by Thaler and Sunstein (2008) is likely to be the most viable route forward regarding sustainability and ethical interest. Sarah Quinlan Cutler (Chapter 10) encourages more ethical information for travellers to be included in guidebooks to influence more responsible behaviour in tourism. The provision of further information to curb consumption was also noted in Wouter Geerts's (Chapter 12) study, which suggested that business travellers would be happy to participate in ethical choices if information was more easily accessible and reinforced by hotel personnel. Sheila Malone's (Chapter 9) chapter argued for tourism organisations to seek out, strengthen and sustain relationships with those customers who demonstrate an interest in ethical consumption.

Concurrent with the inherent tension from the consumer's perspective between relaxation and ethical conduct, Adriana Budeanu and Tareq Emtairah's (Chapter 6) study demonstrates that private operators in the tourism industry are often reluctant to mention environmental and/or socially responsible aspects to tourists, so as not to 'spoil' clients' holidays. Cost was identified to be a barrier to sustainable food consumption as discussed by Kline *et al.* (Chapter 7) who posit tourism as a viable conduit for destinations to develop sustainable food systems for tourism purposes. Lovelock and Lovelock (Chapter 13) address the inequalities inherent in countries where medical tourists seek non-essential health services. Similarly, Michael Hall (Chapter 3) challenges what actually makes alternative quests in tourism unconventional if they are carried out in conventional capitalist economies, suggesting that perhaps such offerings disguise the distressing realities. Along the same lines, Michael Clancy (Chapter 4) points to the paradox of slow travel given travellers, by their very nature, are transient. Wouter Geerts (Chapter 12) presents the business traveller perspective, suggesting that emotional and physical stress act as a barrier against environmental considerations. Furthermore, Clancy (Chapter 4) questions the carrying capacity of slow tourism and its ability to co-exist alongside mass tourism.

The preoccupation in Western societies with consumerism and the accumulation of material goods is proposed as a future challenge for advocates of ethical consumption by Kellee Caton (Chapter 2) and Michael Hall (Chapter 3). Caton (Chapter 2) argues that consuming ethically on holiday presents opportunity to construct an ethical self. This point is echoed by Maria Koleth (Chapter 8) and Karla Boluk and Vania Ranjbar (Chapter 9), who assert that demonstrating an ethical self through volunteering is one way to be marketable (Koleth) or desirable in the presence of significant reference groups (Boluk and Ranjbar). From a psychological and emotional perspective, self-identity is also proposed by Sheila Malone (Chapter 10) and Sarah Quinlan Cutler (Chapter 11). Finally, Clare Weeden (Chapter 14) describes the inherent challenges for those marketing ethical consumption to travellers.

The future of ethical consumption in tourism

Political and economic conditions

The global financial crisis (GFC) has highlighted the absence of ethics in corporate life, as evidenced by widespread public condemnation of corporate greed and executive excess in the finance and banking sector (see for example, the Occupy movement). Indeed, the recession has revealed inferior standards of accountability and a lack of positive leadership across large elements of the corporate world, leading to deep reflection by many about the purpose and value of mass consumerism, with its potentially unsustainable level of consumption in the future. Some have indicated the economic downturn thus has the potential to offer consumers the opportunity to choose the economic system they wish to engage with (Hinton and Goodman 2010: 257). Others argue it has forced consumers to realise that an economy premised on high levels of personal consumption fails on economic and/or environmental grounds (Evans 2011). Seemingly, there has been no better time to explore responsibility in the marketplace, particularly as consumers are currently engaged in a much more reflexive mood due to the poor economic state of the globalised world. Arguably, the 2007–8 GFC will be recognised as the time that spurred a 'responsible turn' in consumptive practices. The shift away from overconsumption may be a consequence of attempting to attain fulfilment, happiness and/or a concern for proving something to one's significant reference groups, and shifting towards reflexive consumption aligned with the goals of sustainability. Thus, the economic moment in time in which we find ourselves could motivate a shift from inwards to outwards, thus becoming a significant opportunity for responsible and ethical consumers in tourism. Such instability may also be the opportune moment to question: how can the tourism industry better engage consumers to make 'responsibility' more personally relevant, and how can tourism businesses better market their sustainable/responsible achievements to encourage ethical consumption in tourism?

Lifestyle and culture of consumption

As described in many of the chapters, the culture of consumption in Western society is problematic and poses a threat to leading more responsible livelihoods supported by ethical consumption. Consuming in an ethical fashion is at risk when the middle classes from the emerging markets of Brazil, Russia, India and China are expected to increase by three billion consumers by 2050 (World Economic Forum 2013). Alternatively, these same individuals could provide a significant opportunity for responsible tourism suppliers, if they act appropriately, to market ethical consumption opportunities to corner this substantial market. Ultimately we need to transform the way people think about ethics in tourism – perhaps we should consider an innovative strategy to remarket the responsible and sustainable discourse to attract alternative demographics, such as business travellers, and younger, engaged, thinking and motivated travellers. Marketing 'values' and/or a redefinition of values in support of sustainable (Hutter et al. 2010) and ethical practice should be considered by business. A focus on acting out one's values and 'doing the right thing' may be the nudge needed to encourage consumers to act ethically in their consumption both at home and abroad. Importantly, the two spheres (home and away) need to be considered in concert if change in consumption is going to be sustainable. Ultimately consumers need to be excited and motivated by ethics and responsibility discourse in order to respond and engage moving forward. Engagement then needs to become a lifestyle for consumers who will want to choose businesses that support ethical consumption goals in the areas of (for example) slow tourism, the 'staycation', fair trade tourism and creative tourism. Likewise, tourism companies need to see the advantage of taking on board ethical goals to enhance their products/services, as well as keeping their consumers informed.

Provision of information

A number of barriers currently exist which significantly impact people's ability to engage in ethical consumption in tourism. Some researchers argue that individual consumers have a limited capacity for behavioural change as they are influenced by 'existing social, technological and market boundaries', and further limited by the extent of their own knowledge (Mont and Plepys 2008: 535). Similarly, a lack of understanding may be the result of the high volume of labels introduced to the market, creating confusion among consumers. Specifically, consumers are unclear of the differences between the promises put forth by labels such as fair trade, ethical and organic (WBCSD 2008). Ultimately, the tourism industry needs to make it easier for people to consume ethically while on holiday; consequently, the provision of information needs to be both explicit and accessible. The lack of information is a current barrier to ethical action in tourism as discussed previously by Hall (Chapter 3), Budeanu and Emtairah (Chapter 6), Quinlan Cutler (Chapter 11) and Geerts (Chapter 12). Specifically it was highlighted that information is not always easy to locate, which can be a disservice to businesses, as such information could be the inclining factor to nudge consumers to support ethical practice.

Comfortable lifestyles/greed and/or selfishness (Young and Dhanda 2013) are described as additional barriers to behavioural change, because although some consumers are aware of the issues, they are disinterested in engaging proactively with possible solutions. For others, perceived financial cost combined with the extra time and effort needed to successfully navigate the ethical market, inhibits positive action. Another factor is the tendency of consumers to follow their peers, thus reflecting the 'tragedy of the commons' (WBCSD 2008). In other words, if an individual's reference group declines to adjust their lifestyle to fit in with responsibility goals, they may not feel compelled to change their own behaviour. Finally, some consumers fail to acknowledge their individual purchasing choices have the power to improve the environment and/or society.

Empowering the consumer

Kline *et al.* (Chapter 7) highlight some of the decisions travellers are faced with in light of where they choose to eat, what to purchase, and where they visit, and note that consumers have the power to make ethical consumption decisions. As such, the onus is typically on the consumer, who is left to research the efforts of companies if they want to support sustainable and responsible business practice. Perhaps tourism businesses need to make it easier for the traveller. Ethical tourism is not likely to be achieved by consumers acting individually ('one holiday at a time' – see Weeden, 2013), but instead the collaborative efforts of many consumers and business ventures that work together. Acknowledging and leveraging the power of consumers is therefore imperative. Consumers have the ability to create business success through the various forms of social media on offer where they can share information about companies, products and services via social networks, to promote sustainable and responsible products, usage, consumption and lifestyles. Consumers are increasingly turning to the internet as a trusted source of peer-generated information: 61 per cent of consumers now consider blogs a reliable source of information, and more than half trust consumer-generated media and branded websites (WCBSD 2008: 21). Perhaps incentivising travellers while they are on sustainable holidays will/could encourage them to engage in social media marketing, which may enhance the profile of some tourism businesses and thus create a virtuous circle.

Conclusion

A few questions have emerged in this book. Are ethical consumers a specific and identifiable market? The answer is *probably*, as more alternative approaches emerge to offer responsible experiences. Since travel is a significant consumption activity its contribution to serious environmental and social problems are apparent. However, responsible and ethical consumption while on holiday is possible and can in fact assist consumers in the continuation of their typical political consumption when away from home, although such decisions may come at a cost, both financially or experientially. This book has put forth a number of arguments to encourage the accessibility of information to travellers, a clarification of and a rebranding of the notion of responsibility in tourism. Furthermore, it has called for the industry to do more to critically engage motivated consumers and to leverage the power of these individuals to support sustainable business operations and consumption.

The contributors to this edited volume suggest individual travellers uphold a range of motivations at different times. Consequently, a tourist could be a keen environmentalist at home but when on holiday they may not demonstrate any interest in, or undertake actions to, protect the environment. Several of the chapters in this book highlight traveller compromise. One example is Chapter 12, where Geerts argues that for some business travellers, environmental and social responsibility falls on the shoulders of the company paying for the business trip, therefore demonstrating that there are a variety of ways in which people are willing to offload their responsibility.

Finally, we intend for our book to supplement what we believe to be a limited understanding of the motivations of ethical tourists and the discourses that influence travellers and their decisions. In particular it has explored how consumers navigate the responsible tourism marketplace and the chapters have provided a rich understanding of the challenges facing those seeking to encourage travellers to become responsible.

References

Evans, D. (2011) 'Consuming Conventions: Sustainable Consumption, Ecological Citizenship and the Worlds of Worth' in *Journal of Rural Studies*, 27: 109–15.

Fennell, D.A. (2008) 'Tourism Ethics Needs More than a Surface Approach' in *Tourism Recreation Research*, 33 (2): 223–4.

Hinton, E. and Goodman, M. (2010) 'Sustainable Consumption: Developments, Considerations and New Directions' in M. Redclift and G. Woodgate (eds), *International Handbook of Environmental Sociology*, Cheltenham: Edward Elgar: 245–61.

Hutter, L., Capozucca, P. and Nayyar, S. (2010) 'A Roadmap for Sustainable Consumption', Deloitte: 46–58. Online, available at: www3.weforum.org/docs/WEF_RedesigningBusinessValue_SustainableConsumption_Report_2010.pdf (accessed 10 December 2012).

Krippendorf, J. (1984) *The Holiday Makers*, London: Heinemann.

Krippendorf, J. (1991) *The Holiday Makers: Understanding the Impact of Leisure and Travel*, Oxford: Butterworth-Heinemann.

Mont, O. and Plepys, A. (2008) 'Sustainable Consumption Progress: Should we be Proud or Alarmed?' in *Journal of Cleaner Production*, 16: 531–7.

Poon, A. (1993) *Tourism, Technology and Competitive Strategies*, Wallingford, Oxfordshire: CAB International.

Poon, A. (2003) 'Competitive Strategies for a "New Tourism"' in C. Cooper (ed.), *Classic Reviews in Tourism*, Clevedon: Channel View Publications: 130–43.

Pritchard, A., Morgan, N. and Ateljevic, I. (2011) 'Hopeful Tourism: A New Transformative perspective' in *Annals of Tourism Research*, 38 (3): 941–63.

Smith, R. and Duffy, M. (2003) *The Ethics of Tourism Development*, London: Routledge.
Stark, J.C. (2002) 'Ethics and Ecotourism: Connections and Conflicts' in *Philosophy & Geography*, 5 (1): 101–13.
Thaler, R. and Sustein, C. (2008) *Nudge Improving Decisions About Health, Wealth and Happiness*, New Haven, CT: Yale University Press.
Weeden, C. (2002) 'Ethical tourism: An Opportunity for Competitive Advantage?' in *Journal of Vacation Marketing*, 8 (2): 141–53.
Weeden, C. (2013) *Responsible Tourist Behaviour*, London: Routledge.
World Business Council for Sustainable Development (WBCSD) (2012) *Business Solutions for a Sustainable World*, Switzerland: Atar Roto Presse.
World Economic Forum (2013) 'Engaging Tomorrow's Consumer'. Online, available at: www3.weforum.org/docs/WEF_RC_EngagingTomorrowsConsumer_Report_2013.pdf (accessed 19 February 2013).
Young, S. and Dhanda, K. (2013) *Sustainability Essentials for Business*, Thousand Oaks, USA: Sage Publications.

Index

ABC (attitude, behaviour, and choice) paradigm 37, 38, 47
ABTA 2, 4
accommodation 8, 83, 90, 92–3; awareness of environmental impact of 94, 97, 98; and carbon dioxide emissions 89; eco-efficiency measures 84; environmentally preferred options (EPOs) 91, 95, 96–7, 99, 242; *see also* hotels
Accor Group 84, 234
action tendency 156
addictive behaviours 234
advertising 21–5
aesthetics 56, 61–6
affect: and medical tourism 221–2; *see also* emotion
air travel 71, 76–7, 85, 229
alternative economies 36–7; alternative hedonism 63–4, 158
alternative tourism 1, 4, 34, 45–8, 122, 225–6, 241
altruism 7, 9, 74, 75, 122–5, 130–8, 145, 149
American Marketing Association 231
Andereck, K.L. 3, 4, 114
animal welfare/rights 5, 25, 72, 107
anthropocentric 71–2, 74, 76
anti-consumption 46, 47
Apollo Hospital Group (AHG) 212–13
appropriate tourism 34
Åre 83, 86–8, 96, 98
aspirational identity 63
Association of American Geographers (AAG) 2
Association of British Travel Agents (ABTA) 2, 4
Ateljevic, I. 114, 195
attentive care 217

attitudes 231 (*see also* consumer behaviour); ABC (attitude, behaviour, and choice) paradigm 37
attitude–behaviour gap 11, 229, 231, 233, 234
aviation 71

backpacker tourism 2, 26
Baobab Travel 230
Barber, B. 21, 22, 24–5, 26
Barnett, C. 5, 37, 44, 59, 60, 123, 138, 210, 211, 213, 214, 218
behavioural economics 37–8, 47
biodiversity loss 70–1, 78
biofuels production 70–1, 78–9
biological drives 22, 23
blogs 10, 240
Boluk, K. 1–15, 58, 107, 123, 134–52, 240–7, 243
boycotts 5, 20, 37, 48, 59, 60
brands/branding 24–5, 63; and consumer loyalty 9, 160, 197–9
Brundtland Report 232, 240
Budeanu, A. 5, 8, 83–103, 153, 188, 229, 242, 244
business travel 10, 188–206, 246; attitudes towards travelling 192–6; emotional/physical stress 193–5, 200, 202, 242; the environment 9, 10, 188–9, 199–203; leisure travellers distinguished 189–90
Butcher, J. 1, 4, 20, 58, 123, 153, 157
buycotts 37, 48, 138

Cafédirect 42
Cape Town Declaration (2002) 227
capitalism 7, 22, 25–6, 36, 45, 46, 47, 73; and conservation 36, 42–3
carbon capture 79

carbon dioxide emissions *see* greenhouse gas (GHG) emissions
carbon neutral positions 34
carbon offsetting 40, 41, 44, 45, 85, 180
Carlisle, S. 3, 26
Cartesian philosophy 74
Caton, K. 4, 6–7, 19–31, 243
certification programs 41, 42, 44, 108, 109, 110
choice, consumer 32, 37, 39–40, 41, 60, 109–10
Christianity 23
citizenship 35–6; citizen-consumer 231–2
Clancy, M. 7, 56–69, 242
Cleverdon, R. 20, 36, 225
climate change 2, 7, 33, 34, 45, 58, 70, 76–7, 199–200, 229, 231
co-creation 107, 114
Cognitive-Affective Model (CAM) 155, 156
commitment: in ethical consumption 9, 160; of volunteer tourists 136
commodification 27, 41; of development 9, 123; of healthcare 207, 209, 210, 214; of indigenous peoples 4
commodity fetishism 7, 26, 35, 41–5
community 7, 35, 72, 78; based tourism 34, 58, 226; capitals framework 114
competition 235
competitive self 125
compromise 2, 6, 157, 241, 242, 246
conflict resolution 27
conscience fatigue 27
conscious consumption 32–3
consequentialist model of ethics 210, 220–1; frameworks 208, 210
conservation 7, 46, 70; and capitalism 36, 42–3; ethic 7, 76–7, 79, 242; wildlife 4
conspicuous consumption 7, 22, 62, 138
consumer behaviour 89; change (social/psychological approaches to 37, 38, 39; systems of provision/institutions approach to 37, 38, 40–1; utilitarian approach to 37, 38)
consumer choice 32, 37, 39–40, 41, 60, 109–10
consumer loyalty 197
consumer product outlets (CPOs) 110–13
consumer rights and safety 209
consumer self 21–5
consumer sovereignty 7, 37, 45, 60
consumerism 6, 20, 243; culture of 21–3, 244; ethical 20–1, 29n2, 33
consumption 37, 56–7, 59–63, 65

consumptive values 61–3, 64
content analysis 10, 171
conventions 111–12
Cooperative Bank Ethical Consumers Market Report 2012 228
corporate sector 243
corporate social responsibility (CSR) 33, 76, 227, 232
cost 33, 245; healthcare 208, 213; and sustainable food consumption 242
Cotlands 139, 143–8
Crane, A. 138, 153, 158, 231, 232
cultural appropriation 4
customer loyalty 9, 160, 197–9

Deale, C.S. 8, 104–21
decision-making 38, 39, 40; consequentialist model of 210, 220–1; development 126–30 (of tourism 1); developmental 75, 78; and emotion 153–60; food consumption 115; medical tourism 211–12, 220–1; and travel guidebook information 10, 169
deep ecology 7, 75–6
deregulation 45, 59
Desertification 70
destination communities: disempowerment of 4; environment 108, 116; impact on 4
Destination Management Organization (DMO) 170–1, 181
development: commodification of 9, 123; as ethical transaction 126–8; self- 123, 124, 127–8, 131, 135, 136–7, 148–9; sustainable 76, 153, 227, 232
development tourism 9, 122–33
developmental decision-making 75, 78
Dickinson, J.E. 56, 57, 58, 85, 86, 97, 229
dignity 227
discourse analysis *see* ethical discourses
Discovery Initiatives 230
Duffy, R. 42, 43, 44, 46, 47, 58

Earth community 75; earth jurisprudence 75; Earth Summit 70
eco-efficiency 84
ecological citizenship 35; scarcity 70
ecological systems theory 114
ecology 72
economic conditions 243; reductionism 26–8; value 70
ecosystem services 43
ecotourism 3–4, 34, 42, 43, 47, 58, 61, 63, 64, 79, 86, 153, 226
education 183–4

Index

egoism 7, 74, 123, 124, 130
emotion 8, 9, 24–5, 153–65; positive and negative 155, 156; *see also* affect
empathy 72
empowerment, consumer 245
Emtairah, T. 8, 83–103, 242, 244
Energy Saving Trust 77
engagement 157, 159
enjoyment 159
enterprise society 124, 127, 130
entrepreneurship 114
environmental awareness 8, 33, 58, 83–4, 84–6, 89–90, 93–5, 97–9, 100; behaviour 8, 199–202; denial strategies 200; economics 71; ethic 7; factual awareness 84, 89, 94; in hotels 10, 200–2; justice 3; low, medium, high 94; *see also* rights
environmental behaviour 85–6; of business travellers 9, 10, 188–9, 199–203; and denial strategies 200; in hotels 200–2
environmental economics 71
environmental ethics 7, 70–80, 242
environmental impact 1–2, 3–4
environmental philosophy 72–7
environmentally preferred options (EPOs) 8, 83, 90, 91, 95–7, 99, 242
ethical blindness 155; complexity of 229; consumerism 20–1, 27, 33, 229, 232; debate 240; decision making 8, 9, 154–6, 158, 228, 241; food 8; goals 244; hedonism 158; holidays 134; identity 6, 8–9; wash/green wash 231
ethical consumption 1, 3, 5–7, 59–61, 63–4, 66, 138, 211, 240–1, 244
ethical discourses, in volunteer tourism 9–11, 134–5, 139–50, 240, 246
ethical self/selving 5, 8, 9, 27, 28, 123, 136–8, 145–6, 149, 153–65, 243
ethical tourism 1, 4, 34, 58, 156–60, 226–7; as an emotional choice 154–6, 159–60; as an enjoyable experience 156–9
ethics 2, 56, 59
European Fair Trade Association (EFTA) 36
evolutionary psychology 22
exclusion 3–4, 62, 63
exclusivity 7, 62
Exodus Travel 230
experiential perspective of consumer behaviour 154, 156–9
experiential value 61–2, 63, 64
exploitation 3, 7

fair trade 3–4, 7, 25, 35–6, 42, 44–6, 59, 61, 104, 107, 111, 114, 138, 226, 244
Fair Trade Labelling Organizations International (FLO) 36
fair trade tourism 4, 5, 47, 135, 138, 226, 244
farming: industrial-scale 106; traditional approaches 106
fast food 57
Fennell, D.A. 1, 2, 3, 4, 153, 157, 158, 240
FINE 36
food access 108–9, 115–6; Alliance 108; citizenship 35–6; festivals 112–13; miles 8, 104, 108; supply chain 8, 105; tourists 110; value chain *see* sustainable food system
food/agricultural tours 113
food systems 8–9, 35, 104–21, 242; consumable product outlets (CPOs) 110–13; consumption/consumer choice 109–10; destination communities 108–9; growth/production 106; harvesting and processing 107; marketing 109, 115; outputs/waste 113, 115; packaging 107, 115; research directions for the future 115–17; stakeholders 8, 105–8, 113–14, 116; transportation and distribution 107, 108
food trucks 112
Fordism 21
Foucault, M. 9, 19, 23, 24, 122, 123, 124–5, 127, 128, 138, 139

gap-year volunteers 125, 126, 136–7
Geerts, W. 10, 188–206, 242, 244, 246
gentle tourism 34
global biological resources 209
global citizen[ship] 126, 134, 137
Global Code of Ethics for Tourism (GCET) 76
global financial crisis (2007–8) 243
global village 72
globalization 57, 72
Goodwin, H. 1, 5, 156, 157, 171, 226, 227
Gössling, S. 35, 58, 84, 234
Gotland 83, 87–8, 96, 98
Green Economy 7, 70, 76
green growth 34, 48
Green Planet Catering 111
green technologies 71
green tourism 34
greenhouse gas (GHG) emissions 34, 58, 71, 85, 89, 108
greenwashing 27, 231

grounded theory 190
guidebooks 9, 10, 169–87; ethical content categories 177, 178, 179; ethical statement themes 173, 174–5; ethical statement types 173, 175–6
guilt 5, 7, 154–7
Guttentag, D. 64, 129, 137, 145, 146, 147

habitual behaviours 5, 39, 63, 86, 211, 214, 234
habitual ethical frames 211–14, 220, 222
Hall, C.M. 3, 7, 32–55, 64, 104, 110, 208, 220, 242, 243, 244
health: of business travellers 195; care consumption 209; of destination communities 108, 116; inequality 209, 213, 221
healthcare *see* medical tourism
hedonism 2, 7, 138, 145, 149, 157–8; alternative 63–4, 158; ethical 158
heritage 4
Higgins-Desbiolles, F. 1, 27, 136, 225
Holden, A. 3, 7, 70–80, 153, 242
hotels 188, 196–9, 200–2; stays (attitudes toward 196–9; and environmental behaviour 200–2; and friction 197, 201; predictability of 198; and social interaction 198–9)
human capital accumulation 9, 125, 127
human rights 5, 25, 27, 72
hypermobility 10, 190, 194

identity construction/expression 6, 8, 9, 21–5, 63, 138
indigenous peoples, commodification of 4
Industrial Revolution 21, 73; industrial-scale farming 106, 113
inequality, healthcare 209, 213, 221
informed decisions 241
information 37, 40, 210, 242, 244–5; deficit 37; overload 39, 228; sources 169–71; in travel guidebooks 9, 10, 169–87, 242
instrumentalism 74–5, 79
intention–behaviour gap 9, 153, 156, 229
International Centre for Responsible Tourism (ICRT) 226, 227
International Centre for Responsible Tourism Canada (ICRT Canada) 171
International Fair Trade Association/International Federation for Alternative Trade (IFAT) 36
International Year of Eco-Tourism (2002) 84

internet 231, 245; and healthcare consumption 209, 211, 212
intrinsic value 74–5, 78
investment, socially responsible 59, 60

just tourism 34

Kalisch, A. 4, 5, 20, 153, 225, 226
Kline, C. 3, 8, 104–21, 242
Knollenberg, W. 8, 104–21
Koleth, M. 9, 122–33, 243
Krippendorf, J. 4, 34, 57, 63, 170, 182, 240, 241

labelling 104, 109, 244
land ethic 73
learned ethical competencies 211, 221
legislation, deep ecology principles in 75, 76
leisure activities 8, 83, 89, 90, 91, 93; awareness of environmental impact of 94, 98; environmentally preferred options (EPOs) 95, 96, 97, 99, 242
lifestyle 25, 244, 245
Lisle, D. 4, 58, 64, 154
LIVE (Low Input Viticulture and Enology) 113
local communities *see* destination communities
localism 8, 35, 59, 114
Lonely Planet India 170
Lonely Planet Peru 170–4, 176–7, 180, 183–4
Lovelock, K. and Lovelock, B. 10, 207–24, 242
lovemarks 24–5
loyalty, customer 9, 160, 197–9

Mabira Rain Forest case study 70, 78–9
Malone, S. 9, 153–65, 242, 243
market failure 71
marketing 10–11, 21–2, 23, 24–5, 160, 243; ethical 229; food tourism 109, 115; PR puffery 230; responsible tourism 225–39; social 7, 33, 47, 48, 232–4; social media 245; sustainable 232; unethical 231; values 244
marketization 35, 38, 48
masculinity 195
mass tourism 2, 3, 4, 57, 124, 125, 225–6; competition 235; tour operators 230
McDonaldization thesis 57, 198
McGehee, N. 112, 135, 137, 138, 143, 146, 148, 149

Index

medical tourism 9–10, 207–24, 242; and affect 221–2; and commodification of healthcare 207, 209, 210; and decision-making processes 211–12, 220–1; emergence and growth of 208–9; and healthcare inequality 208–9, 221; intermediaries and facilitators 217–18; and medical insurance 216–17; and neoliberal healthcare agenda 208; and pain/disability 215–16; and subjective proximity 212, 214, 217, 220, 222; and time 214–15
Meeting Professionals International 111
meetings 111–12
metatourism 34
MINDSPACE report 39
Mintel 153, 154, 227–8, 234–5
modernity 57, 59
mood maintenance theory 156
moral selving 9, 123, 213–14
moral virtue 7
motivations 134, 136, 149, 246
Muir, J. 73–4

national parks 73
nature: intrinsic value of 75, 78; rights of 7, 74, 75–6, 78, 79
neo-classical economic models 7, 37
neoliberal reform 209
neoliberalism 9, 35, 45, 46, 47, 122; and consumer sovereignty 60; and development 123; and healthcare 208, 220; and risk society 59; and self-as-enterprise 122, 124, 128, 129, 130–1
New Puritan movement 4
new tourism 241
niche products 25
non-places 194
nudging 7, 38, 39, 48, 242

optional leisure 89, 93
organic 244
outsourcing 22

packaging 107, 115
participant observation 191
patient-as-consumer 216
patterns of choice 89
Pattullo, P. 153, 154, 155, 158
peace 1, 27
peer groups 245
personal fulfilment, and volunteer tourism 9, 140–3, 148
persuasive messages 175–6, 177, 181–2

Peru 10, 169, 171
planned behaviour, theory of 154
pleasure seeking 157–8
political activism 138; conditions 243; consumption 246
politics, progressive 36–7
Pollan, M. 106
positional goods 7, 62, 63
post-colonialism 136
poverty 70
poverty alleviation 1, 7, 149
Pritchard, A. 1, 194, 241
production 44, 45
progressive politics 36–7, 45
pro-poor tourism 3–4, 7, 36, 47, 58, 64, 135, 226
protected area movement, USA 73
public transport 90, 92, 96, 98
Puritan work ethic 24
Puritanism 4, 32

Quakerism 32
quality of life 108, 115
Quinlan Cutler, S. 10, 169–87, 242, 243, 244

railways 73
Rainbow Tours 230
Ranjbar, V. 5, 9, 134–52, 243
reasoned action, theory of 114
relative choice 89
relative impact intensity of services 90
religious practices 32, 76, 110
respondent bias 88, 90
responsible conduct 240–1
responsible tourism 2–4, 7, 34, 58–9, 153, 177, 179, 180–1, 184, 226, 233, 246; conduct 241; consumer demand 228–30; defining 225–8; marketing 7, 225–39; principles 171, 177, 179, 181, 184
Responsible Tourism Partnership 58
restaurants 111
retail outlets 112
rights 26, 72, 78; environmental 5; human 5; of nature 7, 74–6, 78, 79
risk society 59
Ritchie, J.R.B. 84, 85, 86
Rogerson, C.M. 137, 142, 146
Rorty, R. 19–20, 28
The Rough Guide to Peru 171, 173, 174, 176, 177, 180, 182, 183

sanfter tourism 34

Santos, C.A. 135, 137, 138, 143, 146, 148, 149, 170
satisficing behaviour 37
self: competitive 125; consumer 21–5; ethical 8, 9, 27, 28, 123, 136, 138, 149, 153–65, 243; as multiplicit 23; sense of 136–7, 148
self-development 123, 124, 127–8, 131, 135, 136–7, 148–9
self-as-enterprise 9, 122–31
self-governance 44, 214, 221
selfhood 23–5
self-interest 64, 74
self–other relationship 211, 217
sex tourism 1, 26
Seyfang, G. 35, 37, 39, 40, 46
Sharpley, R. 84, 85, 229
Shaw, D. 5, 32, 33, 37, 44, 46, 136, 154, 160, 229
Sierra Club 73
Simpson, K. 122, 125, 126, 129, 137, 141
Sin, H.L. 123, 135, 136, 137, 138, 141, 146
Singh, S. 135, 136, 137, 142, 149
Singh, T.V. 34, 135, 136, 137, 142, 149
situation ethics 76
slow cities 56, 65
slow food movement 56–7, 61, 65, 104
slow tourism 7, 34, 56–69, 242, 244
"slum" tours 26–7
small-, medium- and micro-sized enterprises (SMMEs) 230
social capital theory 114
social entrepreneurship 114
social justice 4–5, 10, 35, 36, 227, 234
social marketing 7, 33, 47, 48, 232, 233–4
social media 245
social norms 23, 24, 37, 39
social practice theory 229, 231
sociotechnical systems 40, 41, 48
soft tourism 34
soul-hypothesis 23
South Africa 139, 142, 149
South African boycott 20, 60
South American Handbook 171–4, 176–7, 180–3
souvenirs 25–6
Spain, Plan Nacional de Estabilización 74
spectacle 43
Spenceley, A. 1
Stakeholders 6, 8, 225, 227, 240; and food system 8, 105–8, 113–14, 116; stakeholder theory 105, 117
standardization 57, 198

state intervention 35, 37; reduction of 45, 59
status 22, 29n3
staycations 244
steady-state tourism 34
stress, business travel 194–5, 200, 202, 242
subjective proximity, and medical tourism 212, 214, 217, 220, 222
subjectivity 23–4, 219–21
Sugar Company of Uganda (SCOUL) 78, 79
supply chains 35, 46, 231; food 8, 105; theory 105, 117
sustain 117
sustainable development 76, 153, 227, 232; holidays 245; Theory of Planned Behaviour 154; tourism 7–8
sustainable food system: community destination 108–9; consumable product outlets 110–13; consumption/consumer choice while travelling 109–10; definition 104; food tourism 8; growth/production 106; harvesting, process, adding value and packaging 107; marketing 109; outputs/waste 113; transportation and distribution 108
sustainable marketing 232
Sustainable Table 117
sustainable tourism 4, 32–55, 34, 58, 64, 83–103, 104, 153
symbolic or social value 61, 62, 64
sympathy 72

taste 62
Tastes of Portland 113
Tearfund 229, 240
technological solutions to environmental problems 71
Thaler, R.H. 39, 84, 90, 242
Thoreau, H. 73–4
time 215
tour operators 5, 21, 84, 99; and marketing responsible tourism 225–39
Tour Operators' Initiative (TOI) 234
Tourism Concern 5, 154, 227; Ethical Tour Operator Group (ETOG) 234
tourist: addictive behaviour 234; arrivals 34, 70; behaviour 170, 181–2, 184; behaviour change 233; decision-making 10; experience 170–1, 181, 184
tourist consumption 89, 95, 99
trademarks 24
tragedy of the commons 245

254 *Index*

transactional ethics 7
Transcendentalism 32
transformational tourism 1
transportation 8, 83, 84, 89, 92, 96; awareness of environmental impact of 94, 97, 98, 99, 242; environmentally preferred options (EPOs) 90, 95, 97; food 107, 108; *see also* air travel; public transport; railways
Tribes Travel 230
TUI Travel 228, 230, 234, 235

Uganda, Mabira Rain Forest case study 7, 70, 78–9
United Nations 227, 232
United Nations Declaration of Human Rights 72
United Nations Environment Programme (UNEP) 7, 34, 43, 70, 76, 89, 234
United Nations World Tourism Organization (UNWTO) 34, 43, 70, 84, 89, 185, 225, 234; Global Code of Ethics for Tourism (GCET) 76, 171; Responsible Tourist Principles 177, 179, 181, 184
United States Department of Agriculture (USDA) 107, 109
use value 61, 63
utilitarianism 7, 37, 38, 59, 74–5

value-added food products 107, 111
values 41, 84, 86, 154, 158, 159, 160, 244; consumptive 61–3, 64; marketing 244; of nature 73, 78, 79

Varul, M.Z. 123, 138, 144
Veblen, T. 7, 22, 62
vegetarianism/veganism 76
Voluntary Service Overseas (VSO) 240
voluntary simplicity 32, 46, 60
volunteer tourism 4–5, 8–9, 58, 63, 64, 108, 135–40, 243; ethical discourses of 134–5, 139–49 (detachment and reflections on returning home 9, 143–4; distancing discourse 9, 144–6; personal fulfilment 9, 140–3, 148; preferred ways to contribute 9, 146–8); and ethical selving 138; *see also* development tourism

wages 107
waste 113
Watts, M. 36
Wearing, S. 4, 130, 135, 136, 137, 142, 149, 225
Weeden, C. 1–15, 20, 21, 29n1, 153, 240–7
wild law 75
wildlife conservation 4
willingness to pay 85, 86, 188
working conditions 107, 115
World Business Council for Sustainable Development (WBCSD) 244, 245
World Commission on Environment and Development (WECD) 232
World Fair Trade Organization (WFTO) 36
World Tourism Organization 104
Worldwatch Institute 60